WITHDRAWN

MANAGEMENT OF INDUSTRIAL POLLUTANTS BY ANAEROBIC PROCESSES

By Alan W. Obayashi, Ph.D. and Joseph M. Gorgan
Pritzker Department of Environmental Engineering
Illinois Institute of Technology
Chicago, Illinois

INDUSTRIAL WASTE MANAGEMENT SERIES

James W. Patterson, Executive Editor

LEWIS PUBLISHERS, INC.
121 SOUTH MAIN STREET □ P.O. DRAWER 519 □ CHELSEA, MICHIGAN 48118

Library of Congress Cataloguing in Publication Data

Obayashi, Alan W., 1946–
Management of Industrial Pollutants by Anaerobic Processes.

(Industrial Waste Management Series)
Bibliography: p.
Includes index.
1. Sewage sludge digestion. 2. Bacteria Anaerobic. 3. Factory and trade waste—Environmental aspects. I. Gorgan, Joseph M. II. Title. III. Series
TD769.023 1985 628.3′51 84-23378
ISBN 0-87371-001-0

Second Printing, 1985

LEWIS PUBLISHERS, INC.
121 South Main Street, Chelsea, Michigan 48118

PRINTED IN THE UNITED STATES OF AMERICA

PREFACE

This volume, _Management of Industrial Pollutants by Anaerobic Processes_, will be of particular interest to those industrial and environmental engineers who are familiar with fermentation and similar anaerobic process technology and its economic potential, and who seek to identify opportunities for application of this technology in industrial waste management. The volume incorporates a diverse spectrum of information, including an overview of anaerobic technology in its various configurations, as well as their applications for industrial pollutants, and associated toxicity effects. The volume also presents, by way of example, an innovative recovery application to sulfur waste management. This comprehensive volume brings together, for the first time, information on current and potential applications of anaerobic processes in industrial waste management.

Review of this work by the USEPA does not necessarily reflect approval of its contents. Special appreciation is expressed to Mr. William A. Cawley, Deputy Director, USEPA Industrial Environmental Research Laboratory and Chairman of the IWERC Policy Board.

James W. Patterson, Ph.D.
Series Executive Editor and
Director of
The Industrial Waste Elimination
Research Center

SERIES PREFACE

The philosophy of industrial waste management has changed rapidly in recent years, due to regulatory and economic incentives resulting from increasing restrictions on air, water and hazardous wastes pollution, and from escalating costs of pollution abatement. Traditional methods of industrial pollution control often ignore alternative management strategies with significant economic and environmental benefits. Specifically, opportunities for source reduction, by-product recovery, and recycle and recovery for reuse have historically received scant attention. Today, industry is increasingly considering such alternative approaches, and finding many opportunities for their cost effective implementation.

The Industrial Waste Elimination Research Center (IWERC) is a consortium of Illinois Institute of Technology and the University of Notre Dame, established under the auspices of the U.S. Environmental Protection Agency (USEPA), Office of Exploratory Research. It operates under the supervision of a Policy Board, and with advice from its Scientific Advisory Committee and Industry Advisory Council. IWERC has the mission to identify management strategies, and to perform technology evaluation, research, and development on innovative alternative technologies for industrial waste management. As part of its program, the Center commissioned comprehensive studies in three areas:

1. Recovery, recycle and reuse of industrial pollutants;

2. Management of industrial pollutants by anaerobic processes; and

3. Process modifications for pollution source reduction in chemical processing industries.

These studies have yielded state-of-the-art information on opportunities for applications of alternative strategies for industrial waste management. In view of the current widespread and growing interest in such alternative approaches, the results of these studies are being made available, as part of this Industrial Waste Management Series.

James W. Patterson, Ph.D.

LIST OF FIGURES

Figure		Page
1.1	Anaerobic Conversion of Organic Matter to Methane	2
1.2	The Conventional Anaerobic Digestion Process	6
1.3	The Anaerobic Contact Digestion Process	8
1.4	The Anaerobic Upflow Filter Process	10
1.5	The Anaerobic Fluidized Bed Process	13
1.6	The Upflow Anaerobic Sludge Blanket Process	17
1.7	The Modified Clarigester Process	20
1.8	Anaerobic Rotating Biological Contactor [Tait and Friedman, 1980]	22
1.9	Relationship Between pH and Bicarbonate Concentration Near $95^{\circ}F$ [McCarty, 1964]	27
2.1	Proposed Pathway for the Reduction of Benzoate [Whittle and Evans 1975] [Dutton and Evans, 1969]	59
2.2	Proposed Pathway for Degradation of Benzoic Acid to Methane and Carbon Dioxide [Keith, 1978]	62
2.3	Proposed Pathway for the Degradation of Benzoate and Phenol into Methane and Carbon Dioxide [Evans, 1977]	64

Figure		Page

2.4	Proposed Pathway of Phenol and Benzoate Acid Degradation [Neufeld, 1980]........................	65
2.5	Pathway for the Anaerobic Biodegradation of Surfactants [Vath, 1960]...........................	74
2.6	The Structure of DDT and some Related Substances [Kallman & Andrews, 1963]........................	81
2.7	Degradation of Methoxychlor and Heptachlor in Four Soil Types [Castro and Yoshida, 1971]............	83
3.1	General Effect of Compound on Biological Reaction [McCarty, 1964]...	111
3.2	Typical Antagonistic and Synergistic Relationships in a Two Cation System With the Concentration of Cation A Held at a Toxic Level [Kugelman and McCarty, 1965]..........	114
3.3	Response of Unacclimated Methanogens to Nickel [Parkin, *et al*. 1980].......	127
3.4	S.R.T. Response to Toxicant-Nickel, 50 Days at 35°C [Parkin, *et al*. 180].................................	135
3.5	Upper: Normalized Daily Gas Production Versus Time for Methylene Chloride Feed. Calculated Methylene Chloride Accumulation in Digester Versus Time Assuming No Decomposition or Loses (Stuckey, *et al*. 1980)................	137
3.6	Effect of Exposure on Toxicity Response:Nickel [Parkin, *et al*. 1980]................................	139
3.7	Response of Anaerobic Filter to Toxicant Ni^{++} [Parkin, *et al*. 1980]................................	141

Figure		Page
3.8	Response of Anaerobic Filter to Toxicant [Parkin, *et al*. 1980].......	142
3.9	Measured Versus Predicted Methane Production for Toxicant Exposure: Cyanide [Parkin and Speece, 1980].....	148
3.10	Effect of Toxicant Dose Under Schock Loading on Anaerobic Gasification Activity [Pearson, *et al*. 1980]..........................	151
3.11	Effect of Toxicant Dose Under Shock and Acclimation Loading on Anaerobic Gasification Activity [Pearson, *et al*. 1980]................	154
3.12	Calculated Relationship Between Specific Rate of Phenol Utilization and Residual Phenol Concentration Overlayed with Experimental Data Points. [Neufeld, *et al*. 1980]........	154
3.13	Effect of k on Effluent HA_c Concentration [Parkin and Speece, 1980]................................	159
3.14	Effect of K_s on Effluent Acetic Acid Concentration [Parkin and Speece, 1980]................................	160
3.15	Daily Kinetic Evaluation Batch Fed, Continuously Operated Units [Parkin and Speece, 1980].............	161
3.16	Acclimation Effect of Toxicant on V_{max} and K_s [Parkin and Speece, 1980]......	162
3.17	Steady-State Characteristics for Toxicity Kinetic Parameters-Formaldehyde [Parkin and Speece, 1980]................................	164
3.18	Effect of Transient Nature of Toxicity Response on Process Efficiency [Parkin and Speece, 1980].............	165

Figure | Page

3.19 Non-Competitive Inhibition Effect of T_x on Effluent COD [Parkin and Speece, 1980] 167

3.20 Non-Competitive Effect of T_x on Effluent COD [Parkin and Speece, 1980] 168

4.1 The Sulfur Cycle [Cork, 1978] 181

4.2 The Gas Purged Mutualistic System Utilizing Desulfovibrio and Either Chlorobium or Chromatium [Cork and Cusanovich, 1979] 184

4.3 Possible Process Schemes Using Microbial Sulfate Reduction for Treatment of Sulfuric Acid Wastes or Reclamation of Calcium Sulfate Sludges [Middleton and Lawrence, 1977] 185

4.4 Integration of Anaerobic Processes for Sulfur Recovery From a High Sulfate Wastewater 187

4.5 Integration of Anaerobic Processes for Sulfur and Energy Recovery From and Acid Gas Stream 188

LIST OF TABLES

Table		Page
1.1	Advantages Versus Disadvantages of Anaerobic Waste Treatment Processes	4
1.2	Performance of Various Anaerobic Treatment Processes	24,25
2.1	Priority Pollutants Established by the EPA [after Hutton, 1980]	56
2.2	Operational Data of Studies of the Digestion of Food and Agricultural Wastes. [Fannin, *et al*. 1981 and Chynoweth, *et al*. 1979]	70,71
2.3	Results of Screening Studies of Selected Bacteria for Their Capability to Degrade γ-BCH Under Anaerobic Conditions (Jagnow, *et al*. 1977)	72
3.1	Point Source Categories [Keith and Telliard, 1979]	108
3.2	Metal Sulfide Solubility Product Constants [Dickerson, *et al*. 1974]	117
3.3	Comparison of Concentration Levels that Cause Toxicity to Different Anaerobic Treatment Processes	172,173

CONTENTS

1. ANAEROBIC TREATMENT PROCESSES........... 1

Introduction............................ 1
Development of the Anaerobic Process.... 3
Process Control Parameters.............. 23

2. BIODEGRADATION OF INDUSTRIAL ORGANIC COMPOUNDS BY ANAEROBIC PROCESSES...... 47

Introduction............................ 47
Definition of Biodegradation............ 47
Tests Used to Predict Biodegradation.... 48
Biodegradation of Organic Compounds Under Anaerobic Conditions.................. 55
DDT..................................... 79

3. TOXICITY EFFECTS IN ANAEROBIC PROCESSES.............................107

Introduction............................107
Inhibition, Toxicity, Antagonism, and Synergism..............................110
Anaerobic Tests Used for Measuring Toxicity................................123
Toxicity Models.........................147
Listing of Compounds Known to be Toxic to Anaerobic Systems............171

4. MICROBIAL SULFUR RECOVERY-AN EXAMPLE.................................179

Introduction............................179
Microbiology of Sulfur Transformations.........................180
Industrial Applications.................183
Example Process Treatment Schemes.......186
Summary.................................189

CHAPTER 1

ANAEROBIC TREATMENT PROCESSES

INTRODUCTION

The anaerobic degradation of organic matter to methane involves a complex interaction of three groups of bacteria, and is illustrated in Figure 1.1. The first group of bacteria is the fermentative bacteria, which hydrolizes complex long chain organics and ferments them to fatty acids, alcohols and other soluble organics. The second group of bacteria, the acetogenic bacteria has been found to degrade propionate and longer chain fatty acids to acetate, H_2 and CO_2. Presently this is the only known biological degradation pathway for long chain fatty acids and alcohols, since there are no documented cases of methogenic bacteria being isolated which can degrade these organics directly to methane [McInerney, et al. 1979]. As for the third group of bacteria, the methanogens, the substrates for growth which have been determined to date are H_2, acetate, formate and methanol.

McCarty (1964) has suggested that the rate limiting step in anaerobic digestion is the conversion of propionic and acetic acid to methane gas. However, Pfeffer (1979) has indicated that at high solids retention time (greater than 10 days) the rate limiting step in the digestion of sewage sludge at 35 $^{\circ}$C is the hydrolysis of organic solids. In addition, Pfeffer found that cellulose hydrolysis was the rate limiting step in the digestion of municipal solid wastes.

In spite of their present significance and future potential, anaerobic waste treatment processes have not enjoyed a favorable reputation [McCarty, 1964]. This lack of popularity seems to stem partly from ignorance and partly from the many misconceptions held by design engineers concerning

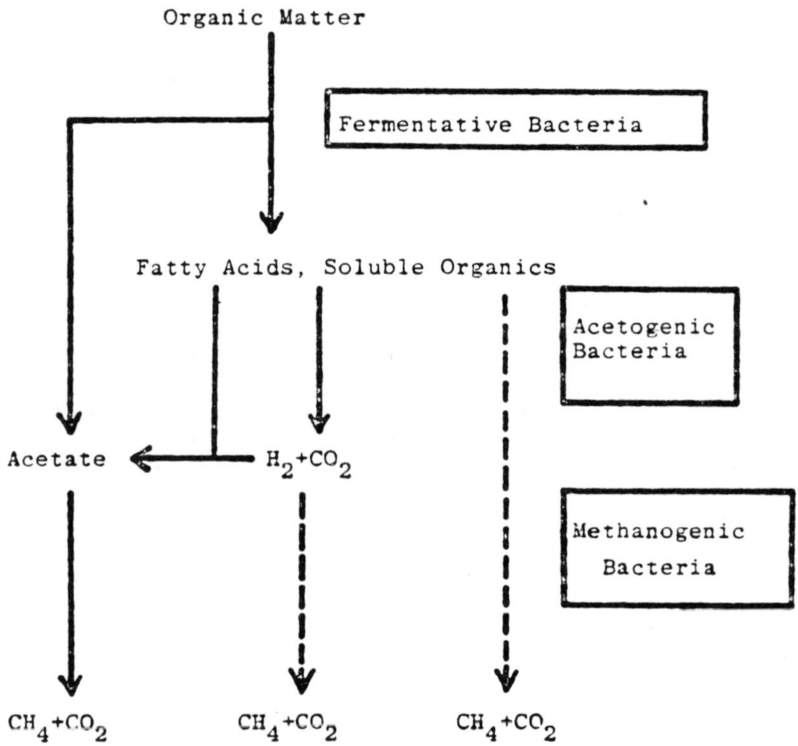

Figure 1.1 Anaerobic Conversion of Organic Matter To Methane

the microbiological and biochemical fundamentals of the anaerobic digestion process. Anaerobic waste treatment is thought of by many as a sensitive process, which is easily upset and difficult to control. The anaerobic digestion process also has a reputation of producing obnoxious odors, as well as requiring long initial startup periods and high temperatures (35°C) for effective waste stabilization. Another possible reason that the anaerobic process has not found general acceptance is that the practical feasibility of direct treatment processes have yet to be proven on specific industrial effluents [Lettinga, et al. 1979].

Anaerobic waste treatment has several important fundamental advantages over aerobic biological treatment processes. The major advantage of anaerobic treatment is the high degree of waste stabilization achieved with very little sludge production (less than 5 percent of the biodegradable organic matter being converted to cell material) [Lettinga, 1979]. Another advantage of the process is that up to 90 percent of the biodegradable fraction can be converted to a usable end-product in the form of methane (CH_4) gas, which can be used to heat the waste stream to give a higher rate of stabilization or to supplement in-plant power requirements. Using direct anaerobic digestion processes, high organic loadings along with short hydraulic retention times (three hours and greater) can be achieved. Table 1.1 [Lettinga, et al 1979] more clearly defines the advantages and disadvantages of anaerobic processes.

In this chapter, two major aspects will be reviewed: first the development of the anaerobic treatment process will be covered and secondly the process control parameters will be discussed. In former sections the processes will be reviewed in a chronological order, i.e., from the simplest to the more high rate processes such as the anaerobic sludge blanket process. In the process control section three areas will be covered. (1) pH and alkalinity, (2) nutrient requirements, and (3) temperature effects on biological processes.

DEVELOPMENT OF THE ANAEROBIC PROCESSES

The full potential of the anaerobic treatment process has not always been realized. Due to the slow growth of methane forming bacteria, relatively

TABLE 1.1

ADVANTAGES VERSUS DISADVANTAGES OF ANAEROBIC WASTE TREATMENT PROCESSES

ADVANTAGES

1. A high degree of waste stabilization is possible at high organic loading rates.

2. Low production of stabilized excess sludge.

3. The excess sludge has good dewatering characteristics.

4. Low nutrient requirements.

5. The use of aeration equipment is not required.

6. Production of a useful-end product in the form of methane.

7. Well adapted sludge can be preserved unfed for a period of one year or more without any appreciable deterioration.

8. May be less sensitive to toxic compounds than aerobic processes.

DISADVANTAGES

1. Start-up of process using digested sewage sludge may last a period of 8-12 weeks.

2. Little practical experience has been gained with the application of the process to the direct treatment of wastewater.

3. Anaerobic digestion is essentially a pretreatment process.

long detention times have been required for the conventional anaerobic process to satisfactorily stabilize organic wastes. For these reasons the conventional anaerobic process (described below) has not been economically feasible for treatment of wastes containing less than 1000 mg/l BOD, because of the large reactor volumes required. Therefore, the conventional treatment process was limited to treatment of high-strength organic wastes such as sewage sludge [Young and McCarty, 1968]. The following sections provide descriptions of the different anaerobic treatment processes that have been developed. Along with each process description, limitations of the process will be pointed out. The anaerobic processes described are the (1) Conventional, (2) Contact, (3) Upflow Filter, (4) Expanded Bed, (5) Sludge Blanker and (6) Anaerobic Rotating Biological Contactor.

Conventional Digestion Process

The conventional anaerobic treatment process consists of a heated reactor containing waste and biological solids (bacteria) responsible for the digestion process. As shown in Figure 1.2, concentrated waste (usually sewage sludge) can be added continuously or periodically (semi-batch operation), where it is mixed with the contents of the reactor. Theoretically, the conventional digester is operated as a once-through, completely mixed, reactor. In this particular mode of operation the hydraulic retention time (HRT) is equal to the solids retention time (SRT). SRT represents the average time the biological solids remain within the reactor. Mathematically, the SRT can be operated as follows:

$$SRT = \frac{M_T}{M_E} \quad (1.1)$$

where: M_T = Total weight of biological solids within the reactor

M_E = Total weight of biological solids leaving the reactor per day.

The fundamental relationship that exists between SRT and process efficiency for biological systems was first described by Lawrence and McCarty

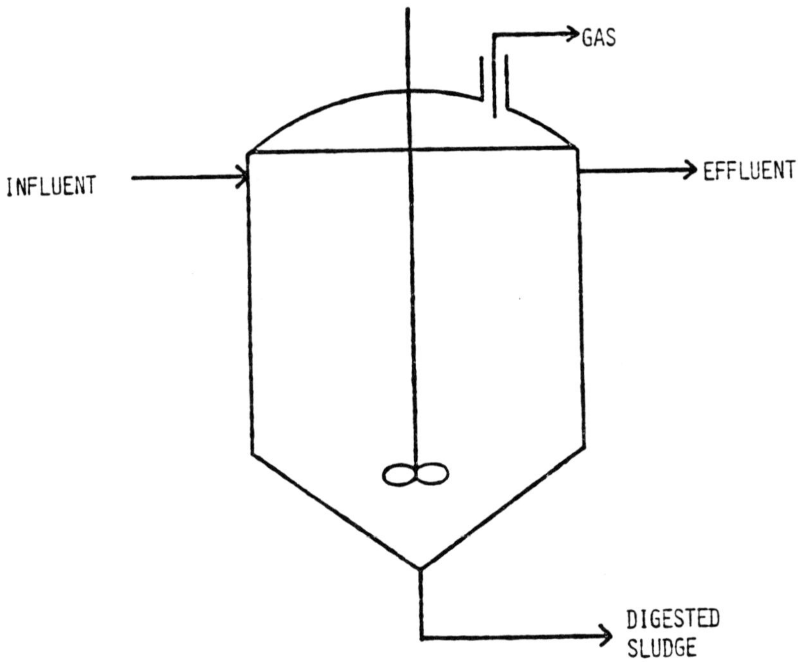

Figure 1.2 The Conventional Anaerobic Digestion Process

(1970). Basically, the required process efficiency is related to the SRT maintained. The SRT maintained is inversely proportional to the net specific growth rate of microorganisms contained in the reactor. This means the longer SRT provided, the longer the microorganisms have to stabilize the available waste and provide a satisfactory population (by reproduction) for further waste stabilization.

By reducing the HRT in the conventional mode reactor, the quantity of biological solids within the reactor is also decreased as the solids escape with the effluent. The limiting HRT is reached when the bacteria are removed from the reactor faster than they can grow. Methanogenic bacteria are slow growers and are considered the rate-limiting component in the anaerobic digestion process. Being slow growers, methanogenic bacteria require long SRT (and thus long HRT) values, with the limiting value ranging from 3 to 5 days at 35°C. To ensure proper control and reliable treatment, detention times usually range from 10 to 30 days at 35°C [McCarty, 1964].

The first major improvement of the anaerobic processes to be made was related to separation of the SRT from HRT. By separating the SRT from the HRT, smaller reactor volumes and thus higher loadings could be treated. Therefore, industrial wastes which were relatively soluble would be treated at lower HRT's, while maintaining the required SRT necessary for the methanogenic bacteria.

Anaerobic Contact Process

The first anaerobic process developed which separated the SRT from the HRT was the anaerobic contact process (also known as the activated anaerobic sludge process). As shown in Figure 1.3, the anaerobic contact process consists of a reactor and a settling basin, and is similar in configuration and operation to the activated sludge process. Much of the pioneering work on this process was conducted by Schroepfer and his colleagues in the 1950's [Schroepfer, *et al*. 1950]. In this process the waste is fed to the reactor, which contains a high concentration of biological solids. Like the activated sludge process, a high concentration of solids is maintained in the reactor by

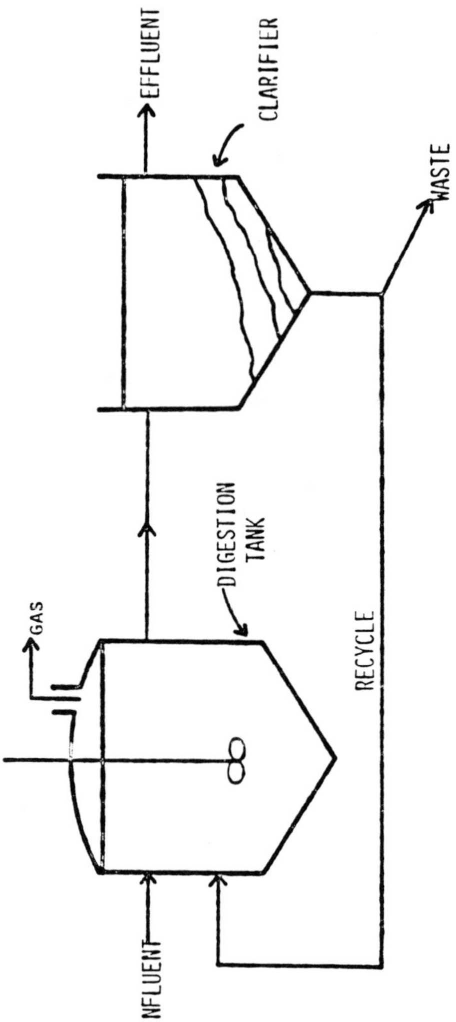

Figure 1.3 The Anaerobic Contact Digestion Process

means of a settling basin, and are returned from there to the reactor. The solids concentration is controlled by solids wastage and by the thickening ability of the sludge.

The key to this process is control of the SRT by the amount of solids wasted. This means that the HRT can be reduced if an adequate SRT is maintained. Many laboratory and pilot-plant studies were conducted successfully using the anaerobic contact process in treating kiering and slaughterhouse [Pettet, _et al_. 1959], citrus [McNary, _et al_. 1952], and synthetic [Dague, _et al_. 1966] wastes ranging from 2000 to 6000 mg/l BOD. A full scale anaerobic contact process (30-35°C) treating a meat packing waste at a loading of 2.5 kg COD/m^3/day, a HRT of 13.3 hrs., and a 90 percent removal of COD was reported by Steffen and Bedker (1966). On the basis of the reported data, the SRT is about 13 days.

The basic disadvantage in the contact process is the requirement of an extra unit process and the associated control problems similar to what occurs in the activated sludge process. Furthermore, in the treatment of a meat packing waste [Steffen and Bedker, 1966; Schroepfer, _et al_. 1955] a degasifier was required in order to aid in the settling of the solids.

Anaerobic Upflow Filter

The concept of biological growth retained on support medium or packing material is not new in the field of wastewater treatment. Aerobic trickling filters use the fixed bed principle as the basis for their operation. Until recently, this approach had not been previously used as an anaerobic process.

Young and McCarty (1968) were the first to recognize the need for an ideal anaerobic waste treatment process that would be able to retain long SRT's independent of waste flow, in order to increase organic loadings. In 1963, they began work which eventually led to the development of the anaerobic upflow filter process. As shown in Figure 1.4, the system consists of an upflow reactor which contains (in their case), 2-2 1/2 inch rock media. As the waste is uniformly distributed (to ensure optimum exposure to

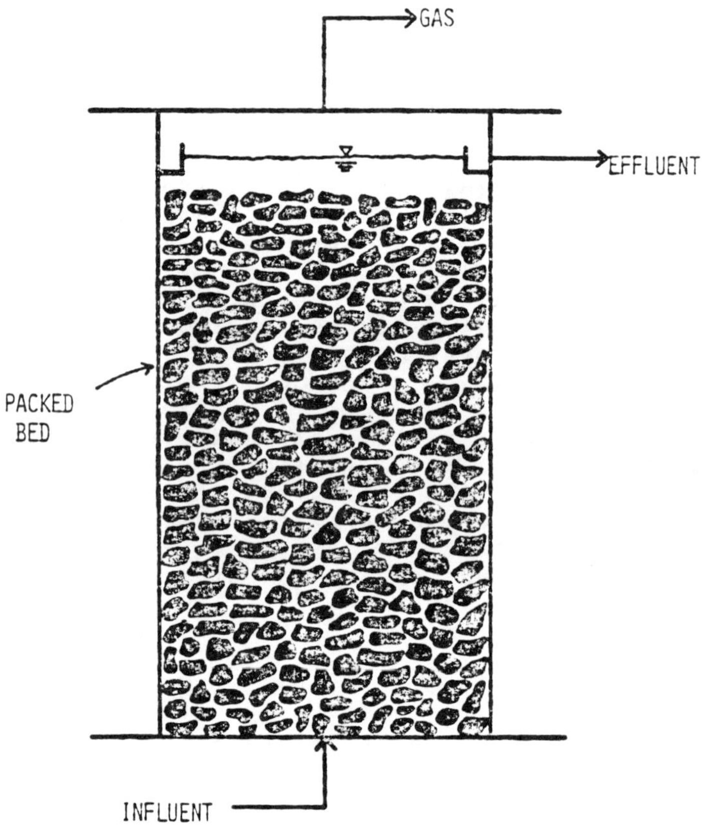

Figure 1.4 The Anaerobic Upflow Filter Process

biological solids) through the filter media, the biological solids develop on the media surfaces or become trapped in the voids between the media. This prevents their release into the effluent stream. The packed filter media, while retaining biological solids, also provides a mechanism for separating the solids and the gas produced in the digestion process.

The anaerobic filter represented a significant advance in anaerobic waste treatment, since the filter can trap and maintain a high concentration of biological solids. By trapping these solids, long SRT′s could be obtained at large waste flows, necessary to anaerobically treat low strength wastes at nominal temperatures economically.

There are several well documented laboratory, pilot-plant and full-scale studies that have been carried out to demonstrate the viability of the process. Young and McCarty (1968) were the first to complete a laboratory study demonstrating that the anaerobic filter was a feasible process. Their study was carried out using a volatile acid and a protein-carbohydrate waste, each applied to a different reactor. They concluded that for both wastes, loadings ranging from 0.42 to 3.4 kg $COD/m^3/day$ (at 25^oC) could be treated achieving COD reductions of 60 to 98 percent.

Jennet and Dennis (1975) conducted laboratory studies on the feasibility of treating a pharmaceutical waste containing 95 percent methanol (on a COD basis) and a small fraction of toulene. They were able to achieve a loading of 3.5 kg $COD/m^3/day$ (at 37^oC) along with a 97 percent COD reduction of the incoming waste. Sachs, *et al.* (1978) conducted laboratory studies on the treatability of synthesized organic chemical pharmaceutical wastes, and achieved an 80 percent COD reduction at a loading of 0.56 kg $COD/m^3/day$ (at 35^oC) and 36-hour hydraulic retention time.

Obayashi and Roshanravan (1980) successfully treated rendering plant wastes at pilot-plant scale using the anaerobic filter process. The pilot-plant was able to achieve a COD reduction of 70 percent at a loading of 2.0 kg $COD/m^3/day$ (at 35^oC) and a hydraulic retention time of 36 hours. The Celanese Company located in Corpus Christi, Texas

[Witt, et al. 1979] successfully installed a full-scale anaerobic filter to treat guar bean waste. The filter was originally designed to handle a loading of 1.0 lb. COD/ft /day at a hydraulic retention time of 1 day and obtain a 65 percent COD reduction.

In 1976, DeWalle and Chian (1976) conducted experiments using an upflow anaerobic filter to demonstrate whether or not a completely mixed flow regime within the filter reactor could successfully treat low pH incoming wastes. Unlike the previous filter studies, which used a plug flow regime throughout the reactor, DeWalle and Chian were able to produce a complete-mix flow regime by having a relatively high recycle ratio.

In their study they used a landfill leachate, which was a very high strength wastewater and contained free volatile fatty acids and complex high molecular weight carbohydrate-like organics, yielding a chemical oxygen demand (COD) of 54,000 mg/l and a pH of 5.4. The fatty acids represented 45 percent of the waste COD.

Taking advantage of the fact that the effluent had sufficient alkalinity, the effluent was recycled in an amount sufficient to raise the influent pH (to about 7.0) without the need of chemical addition. DeWalle and Chian concluded that a high strength wastewater with an acid pH could be successfully treated using a completely mixed anaerobic filter, and by doing so negating the additions of costly buffer solutions. They were also able to obtain a maximum substrate removal rate of 7.0 kg COD/m^3/day with 89 percent removal efficiency.

Anaerobic Attached Film Expanded Bed Process

Another anaerobic process which relies on the development of biomass on the surfaces of a media is an expanded bed, upflow reactor (Figure 1.5). The primary concept of the process consists of passing wastewater up through a bed of inert sand sized particles at sufficient velocities to fluidize and partially expand the sand bed. The inert particles act as a support surface for the growth of large amounts of attached microorganisms [Jeris, et al]. The key to the process is the enormous surface area provided by the relatively

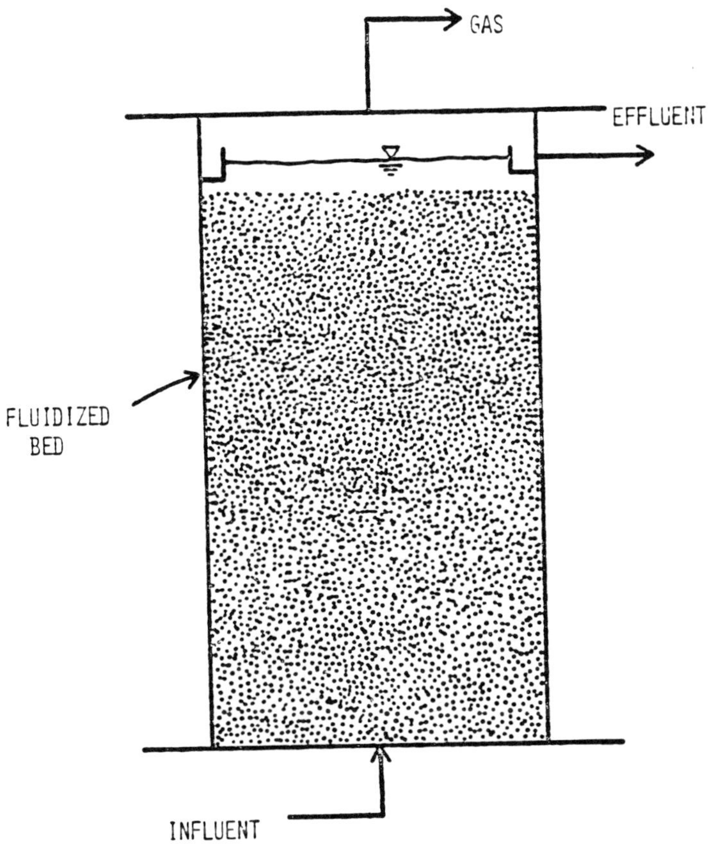

Figure 1.5 The Anaerobic Fluidized Bed Process

small particles for microbial growth and, thus, waste stabilization. A study by Switzenbaum and Jewell (1978) reported solids concentration within the expanded bed exceeding 40,000 mg/l, where 95 percent of the total mass was observed to be attached to the media. In another study Jewell (1980) found the expanded bed process had the ability to accumulate a concentration of biomass exceeding 30,000 mg/l. The inert media also plays the role of a filter, by entrapping small particles which is essential in maintaining high biomass concentrations at relatively high flow rates.

In order to avoid the higher flow rates required (in order to reduce energy consumption) to maintain particles in a fluid state, the upper flow velocities are limited to those required to achieve a 10 to 20 percent bed expansion. In order to further reduce process energy inputs, the density of particles should be as light as possible, and still enable rapid separation from the wastewater flow. Thus, the low flow-through velocities and small bed volume expansion represent the only differences between the expanded bed and fluidized bed processes [Jewell, 1980].

Originally the expanded bed process was developed as a fluidized bed and primarily as an anaerobic waste treatment process. With the excessive clogging experienced in the fluidized bed due to massive biological growth under aerobic conditions, smaller velocities and particles were adopted, and the expanded bed evolved. The expanded bed process, can be used for BOD reduction, nitrification, or dentrification [Jeris, et al. 1977].

Jewell and Switzenaum (1978) presented results from a study that showed the expanded bed process under anaerobic conditions was effective for treating low strength organic substrates (COD's less than 600 mg/l) while operating at low temperatures (10 to 30°C) and at organic loadings of 0.8 to 4.8 kg COD/m^3/day and at small hydraulic retention times of 0.33 to 6 hours. The waste used was synthetic and consisted of glucose, yeast extract and several inorganic nutrients. COD removals of 80 percent or better were achieved for all the conditions tested.

Jewell, *et al*. (1981), reported on a study in which the principal objective was to determine whether it was feasible to treat domestic sewage using the anaerobic attached film expanded bed reactor at a temperature of 20°C. The influent waste was fed continuously to the reactor, and had an average COD value of 186 mg/l (range of 88 to 306 mg/l). The influent suspended solids averaged 86 mg/l and ranged from 40 to 186 mg/l. The hydraulic retention times during the study varied from 5 to 180 minutes. The duration of the study lasted a total of 250 days, with the first 50 days being used for start-up.

The results of the study showed that the attached film expanded bed reactor was able to reduce the influent sewage to less than 40 mg/l as total COD and 5 mg/l suspended solids, at hydraulic retention times of less than 30 minutes.

At most hydraulic retention times (up to 2 hours and greater), COD removals as high as 80 percent were achieved. Maximum organic removal efficiencies (80 percent) were achieved at organic volumetric loadings as high as 4.0 kg COD/m^3 per day. As noted by Jewell, *et al*. (1981), this is the first time that anaerobic treatment of dilute organic wastewaters at a relatively low temperature had been shown to be feasible.

Upflow Anaerobic Sludge Blanket Process

One of the more interesting new processes is the UASB, or upflow anaerobic sludge blanket process (shown in Figure 1.6), which was developed by Lettinga and his co-workers in Holland in the early 1970's [Lettinga, 1979; Lettinga, *et al*. 1980]. The key to the process was the discovery that anaerobic sludge inherently has superior flocculation and settling characteristics, provided the physical and chemical conditions for sludge flocculation are favorable. When these conditions are met, a high solids retention time at high loadings can be achieved, with separation of the gas from the sludge solids [Lettinga, 1980].

Basically, the reactor consists of 3 distinct zones; the sludge bed, sludge blanket, and settling/gas separation zone. The waste is initially fed into the reactor. Like any of the other upflow processes previously described, care must be taken to ensure that the incoming waste flow is uniformly distributed across the entire cross-sectional area of the reactor in order to promote efficient substrate utilizatiohn by the microbial population, and to prevent short-circuiting through the sludge bed zone. The waste flow rate passes through the sludge bed zone. The waste flow first passes through the sludge bed zone (see Figure 1.6) which is formed by the settling and thickening properties of the anaerobic (active biological solids) sludge. Waste stabilization occurs as the waste passes up through the sludge bed, with solids concentrations in the sludge bed having been reported to reach as high as 100-150 g/l [Pette, *et al*. 1979]. These high concentrations in the bed were due mainly to the development of a highly developed granular sludge (consisting of active biomass or bacteria) which has superior settling properties [Lettinga and Zeeuw, 1980]. Mechanical mixing of any kind in the sludge bed zone should be kept to a minumum (except during initial start-up periods), since it is possible to erode some of the granular shaped sludge particles.

The sludge bed zone has been described as a perfectly mixed region [Heertjes and vander Meer, 1978] which can actually be divided into smaller subregions. The first subregion encountered in the sludge bed is the area around the influent ports, which is considered to be a perfectly mixed region. The rest of the sludge bed (except for dead space regions), is considered a transition region between the initial bed zone and the sludge blanket zone. Gas production by the biological solids results in gas bubbles, which produce mixing throughout the bed and eventually eliminating any initial dead space regions. One reason for considering the sludge bed as less than a pefectly mixed zone is that some of the influent bypasses the sludge bed zone via canals formed in the bed. The sludge bed zone is responsible for 80 to 90 percent of the waste degradation occurring in the reactor while occupying roughly only 30 percent of the reactor volume.

The next zone encountered by the waste stream is the sludge blanket section, which occupies about 70 percent of the total reactor volume and contains solid concentrations an order of magnitude lower than the sludge bed. The sludge blanket zone is considered ideally mixed and consists of highly

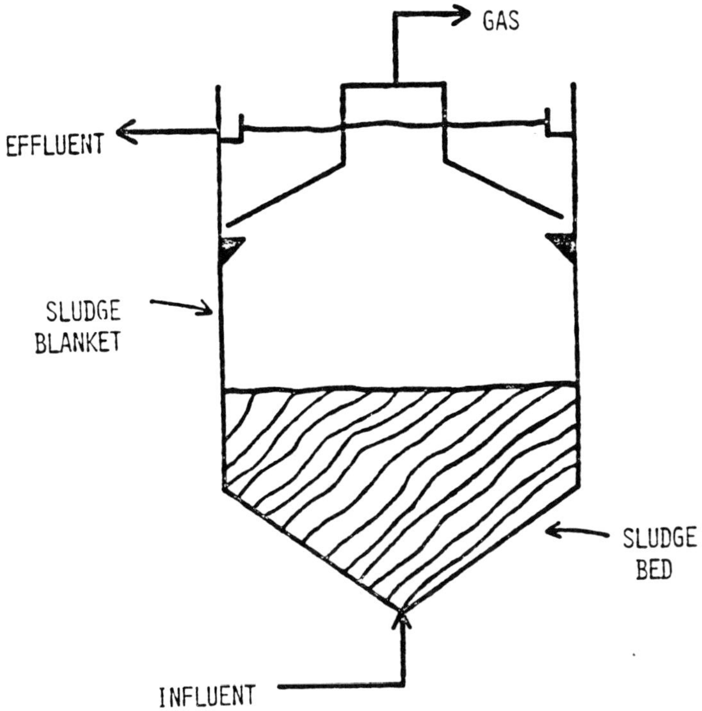

Figure 1.6 The Upflow Anaerobic Sludge Blanket Process

flocculated sludge. The free rising gas bubbles in the blanket (originating from the sludge bed) cause mixing, and because the liquid residence time is very long compared with the time the gas bubbles need to reach the reactor top, very good mixing conditions exist [Heertjes and van der Meer, 1978].

As gas bubbles rise through the sludge bed and blanket, biomass particles become attached and are carried up along with the gas. If long SRT values are to be achieved, these rising solids particles must be separated from the gas bubbles. The separation process is done by a combination solids settler/gas collector device, at the top of the reactor. The main purpose of the settling device is to drive the rising gas and solids particles in toward the gas collector. At this point, a swirling action occurs as the gas goes into the collector, and the solids settle back down into the reactor. The effluent flows through the available space between the gas collector and settler, up toward the effluent weir. Solids that do escape with the effluent, in most cases, settle back down on the settler and drop back down into the reactor.

Heertjes and van der Meer (1978) varied several parameters such as gas production rate, sludge bed height, and solids settler/gas collector design to determine if any of the variations affected the fluid flow through the reactor. They concluded that increases in gas production rates and changes in solids/settler gas collector designs had no effect on fluid flow. They did find, however, that increasing the height of the sludge bed increased the amount of flow that bypassed the bed. This was considered to occur due to the increased thickening of the sludge bed, thus increasing the difference between densities of the influent and sludge mass.

In the past 10 years (1970-1980) a considerable amount of research and development on the UASB has taken place at all levels (laboratory, pilot-plant and full-scale). Literature from Holland has reported that skimmed milk [Lettinga and Vinken, 1980], potato [Lettinga, _et al_. 1980], methanolic [Lettinga, _et al_. 1979], liquid sugar and fatty acids [Heertjes, _et al_. 1979], and sugar-beet wastes could be treated successfully with the UASB process. In pilot studies, organic loadings of 15 to 40 kg COD/m^3/day at liquid residence times of 3-8 hours successfully treated high strength wastewaters. An interesting point in the above literature was the construction of a 200 and then a 800 m full-scale UASB reactor at the Halfweg factory in Holland, for treating sugar beet wastewater [Pette, 1979]. Originally, the 200/m^3 reactor was constructed to treat approxinmately one quarter of the plant's wasteweater flow. The reactor was able to treat the sugar-beet waste at organic loadings as high as 16 kg COD/m^3/day (influent COD concentrations equal to 4200 mg/l), and at a hydraulic retention tinme of 7.1 hours. Removals as high as 90 percent were achieved, and the decision was made to construct an 800 m^3 reactor at the plant site.

The new larger reactor was designed to handle the sugar-beet plant's total daily wastewater output. With operational data from 1978 [Pette, _et al_. 1979], it was found that the 800 m^3 reactor was capable of treating organic loadings as high as 10 kg COD/m^3/day at a hydraulic retention time of 4 hours, with a purification efficiency of 80 percent.

Modified "Clarigester" Process

A process similar to the upflow blanket process is called a modified clarigester (Figure 1.7). A regular clarigester consists of a clarifier over an unheated digestion tank and is similar to an Imhoff tank. The clarigester is modified by changing the influent feed from the clarifier section to the lower digester compartment. By making this change, the flow pattern is similar to the upflow regime of the sludge blanket process. Through the influent inlets, an upflow effect is created throughout the digester compartment, displacing the mixed sludge liquor up through the reactor into the clarifier portion where the gas, solids and supernatant are separated. Any solids entering the clarifier are returned back into the digester portion of the reactor, by means of a scraper.

Using the modified clarigester, Stander (1967) performed successful pilot-plant and full-scale studies in the treatment of a wine distillery waste achieving a 97.3 percent reduction (22,000 mg/l influent COD) at loadings of 32.2 kg COD/m^3/day with a hydraulic retention time of 8 days. Wastes such as maize-starch and compressed yeast waste were also

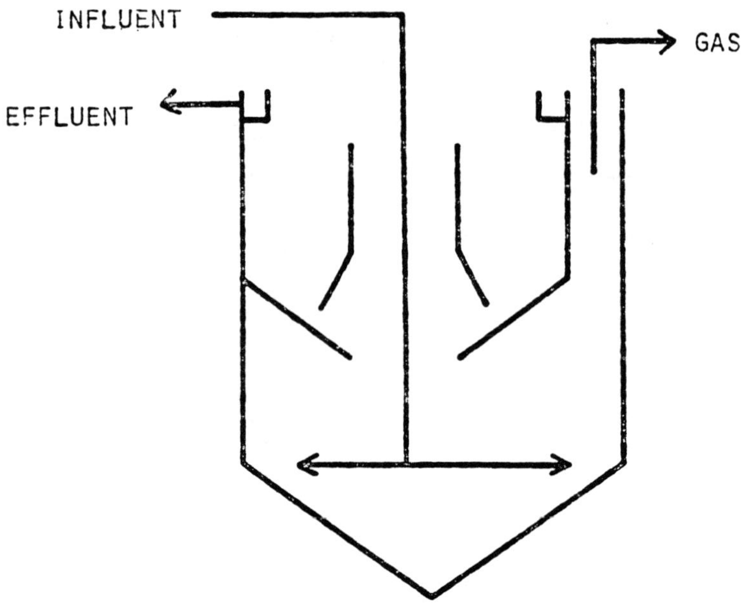

Figure 1.7 The Modified Clarigester Process

treated successfully in pilot-plant and full-scale systems [Ross, 1980]. For the maize-starch waste, a 93 percent reduction was achieved at a loading of 2.4 kg COD/m^3/day (at 24°C) and a hydraulic retention time of 3.3 days. Compressed yeast waste was treated at a loading of 4.0 kg COD/m^3/day (at 35°C) and an 8 hour hydraulic retention time while achieving a 69.5 percent reduction.

Anaerobic Rotating Biological Contactor Process

Another anaerobic attached growth treatment process currently being developed [Tait and Friedman, 1981] to treat high-strength industrial wastewater is the anaerobic rotating biological contactor (AnRBC). Originally, the rotating biological contactor (RBC) was developed as an alternate aerobic treatment process. The AnRBC process has the ability to take advantage of the fixed-film, horizontal flow capabilities of the aerobic RBC process, plus the high-strength, carbonaceous degradation potential of anaerobic systems.

The four-stage AnRBC reactor used by Tait and Friedman is shown in Figure 1.8 and can be described as a series of disks mounted on a horizontal shaft in an enclosed cylindrical tank. All of the disks are partially submerged in the wastewater and are continuously rotated in order to provide mixing, and facilitate product gas transfer to the anoxic atmosphere above the water surface. Each state is separated by a baffle containing ports, to prevent short circuiting and permit flow passage from stage to stage. Biomass growth on the disks ensure cell retention in the reactor.

Tait and Friedman in their studies using the AnRBC process, used a synthetic waste (containing sucrose as the sole carbon source) for the purpose of evaluating the system at several hydraulic and volumetric loadings to determine under what conditions both anaerobic nonmethanogenic and methanogenic microorganisms can develop on the rotating disk surfaces. They were able to run the rotating contactor unit at volumetric loadings ranging from 4.1 to 32.6 kg COD/m^3/day (hydraulic retention times ranging from 17.5 to 2.19 hrs.), with corresponding process efficiencies of 96 percent to 46

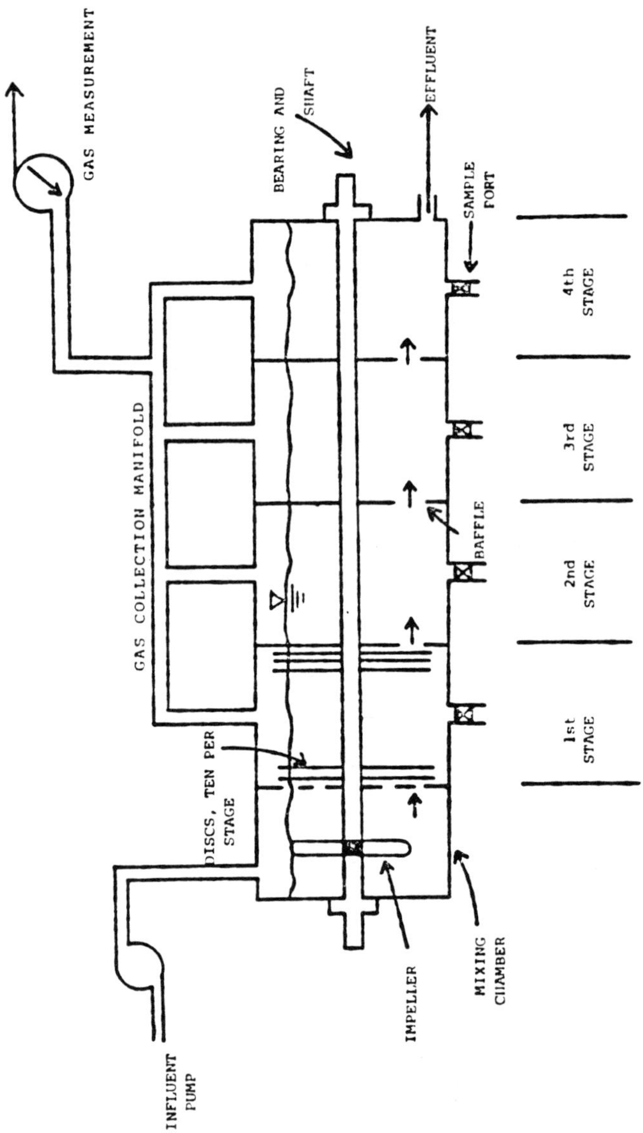

Figure 1.8 Anaerobic Rotating Biological Contactor (Tait and Friedman, 1980)

percent, at a temperature of 35°C. The lower process efficiencies were obtained at the lower hydraulic retention times.

In Table 1.2, a comparison is made of the performance of the various anaerobic systems described previously. As can be seen, most of the reported work consists of laboratory and pilot-plant scale studies, with few full-scale studies. Of the reported work, the highest loading rates are with the sludge blanket process, where loading rates of 10-15 kg COD/m^3/day have been attained in full-scale systems.

PROCESS CONTROL PARAMETERS

In this section, the following three anaerobic process control parameters will be addressed, (1) pH and alkalinity, (2) nutrient requirements, and (3) the effect of temperature on biological processes. Inhibition and toxicity effects are addressed in Chapter 3.

pH and Alkalinity

The pH of the waste and associated alkalinity is an important parameter. In particular, the methanobacteria are pH sensitive, and have an optimum pH range of 6.6 to 7.6 [McCarty, 1964]. Beyond these pH limits, digestion can proceed but with less efficiency. At a pH below 6.2, the efficiency drops off rapidly, and the acidic conditions produced can become inhibiting to the methane bacteria. For this reason it is important that the pH not be allowed to drop below this value for a significant period of time.

Since one of the primary substrates of the methane forming bacteria is acetic acid, any sudden environnmental change usually results in a pH drop which, if not controlled, results in process failure. When a digester is in balance, the acetogenic and methane bacteria use the acid intermediates as rapidly as they are formed. If the methane bacteria are not present in suitable numbers or are being inhibited by unfavorable conditions, they will not use the acids as rapidly as as they are produced which results in an increase

Table 1.2. Performance of Various Anaerobic Treatment Processes

Process	Type of Waste	Loading ($Kg/M^3/d$)	HRT (hrs)	Temp. (^{o}C)	Removal	Scale	Reference
Conventional	Cane Sugar	0.62 (COD)	96	37	60%	Pilot	Bhaskaran and Chakrabarty, 1966
	Yeast	1.7 (BOD)	93	35	70%	Full	Pettet, *et al*. 1959
Contact	Meat Packing	3.2 (BOD)	12	30	95%	Pilot	Schroepfer, *et al*. 1955
	Meat Packing	2.5 (BOD)	13.3	35	95%	Full	Steffen and Bedker, 1961
	Keiring	0.085(BOD)	62.4	30	59%	Pilot	Pette, *et al*. 1959
	Slaughter House	3.5 (BOD)	12.7	35	95.7%	Pilot	" " " "
	Citrus	3.40 (BOD)	32	34	87%	Lab	McNary, *et al*. 1962
Modified Clarigester	Distillery	3.2 (COD)	168	35	99.6%	Full	Stander, 1967
	Compressed Yeast	4.0 (COD)	192	35	69.5%	Full	Ross, 1980
	Maize-Starch	2.4 (COD)	79.2	24	93.0%	Full	" "
Upflow Filter	Synthetic	1.0 (COD)	--	25	90%	Lab	Young and McCarty, 1961
	Pharmaceutical	3.5 (COD)	48	35	98%	Lab	Jennet and Dennis, 1975
	Synthesized - Pharmaceutical	0.56 (COD)	36	35	80%	Lab	Sach, *et al*. 1979
	Guar Gum	7.4 (COD)	24	37	60%	Full	Witt, *et al*. 1979
	Rendering	2.0 (COD)	36	35	70%	Pilot	Obayashi, 1980

Table 1.2. Continued

Process	Type of Waste	Loading ($Kg/M^3/d$)	HRT (hrs)	Temp. (^oC)	Removal	Scale	Reference
	Landfill Leachate	7.0 (COD)	--	25	89%	Pilot	DeWalle and Chian, 1976
Expanded Bed	Synthetic	0.8-4.8 (COD)	0.33-6	10-30	80%	Lab	Switzenbaum and Jewell, 1978
	Domestic Wastewater	4.0 (COD)	--	20	80%	Lab	Jewell, et al. 1981
Sludge Blanket	Skimmed Milk	7.1 (COD)	5.3	30	90%	Lab	Lettinga and Vinken, 1980
	Sauerkraut	8-9 (COD)	--	--	90%	Lab	" " " "
	Potato	25-45(COD)	4	35	93%	Pilot	Lettinga, et al. 1980
	Sugar	22.5 (COD)	6	30	94%	Pilot	" " " "
	Champagne	15 (COD)	6.8	30	91%	Full	" " " "
	Sugar-Beet	10 (COD)	4	35	80%	Full	Petto, 1979
	Brewery	95 (COD)	--	--	83%	Pilot	J. Oat Co., 1981
	Potato	10 (COD)	--	--	90%	Pilot	" " " "
Biological Contactor	Synthetic	4.1-32.6 (COD)	2.19-17.5	35	46-96%	Lab	Tait and Friedman, 1980

in the volatile acid concentration. Thus, an increase in acid concentration indicates that the methane formers are not in balance with the acid formers.

In anaerobic digestion, the process is carried out in the pH range of 6 to 8 where the major chemical system controlling pH is the carbon dioxide-bicarbonate equilibrium. The system is related to pH or the hydrogen ion concentration through the following equilibrium [McCarty, 1964]:

$$H^+ = K_T \frac{H_2CO_3}{HCO_3} \tag{1.2}$$

where:

K_T = ionization constant for carbonic acid

H_2CO_3 = carbonic acid concentration, related to the percentage of carbon dioxide in the digester gas, and

HCO_3^- = bicarbonate ion concentration, forming part of the total alkalinity

The bicarbonate ion concentration or bicarbonate alkalinity is approximately equivalent to the total alkalinity for most wastes, when the volatile acid concentration is low. Basically, alkalinity represents the total amount of acid that can be neutralized by a base when added to the system. When volatile acids begin to increase in concentration, the pH is buffered by the bicarbonate alkalinity.

As shown in Figure 1.9 [McCarty, 1964] and described by equation 1.2, the concentration of carbon dioxide gas and the bicarbonate alkalinity are related. Low alkalinity values in an anaerobic reactor do not allow much of a safety factor for increases in volatile acids. It is, therefore, desirable to have a bicarbonate alkalinity value in the range of 2500 to 5000 mg/l, in order to provide a buffer capacity to handle volatile acid increases with a minimal decrease in pH. If increases in the volatile acid concentration drop the alkalinity to low concentrations and a serious pH drops occurs, supplemental bicarbonate alkalinity should be added.

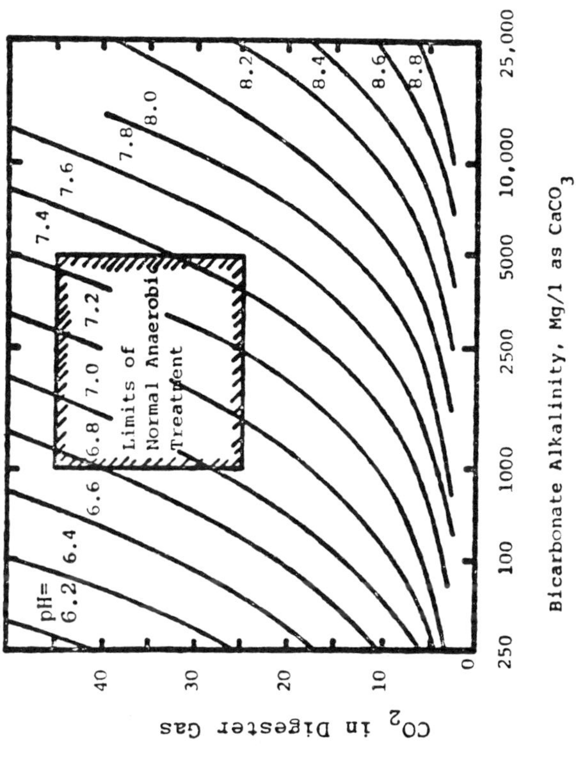

Figure 1.9 Relationship Between pH and Bicarbonate Concentration Near 95°F (McCarty, 1964)

If sufficient amounts of natural alkalinity are not present in the waste stream, then the alkalinity can be controlled by reducing the feed rate, which will allow the volatile acids already present to be utilized, and thus decrease its concentration. Natural alkalinity in waste streams are due to the presence of the hydroxides, carbonates, and bicarbonates of elements such as calcium, magnesium, sodium or ammonium. Of these, calcium and magnesium bicarbonates are the most common [Chow, et al. 1972). An alternate solution in controlling the alkalinity and pH can be accomplished by adding alkaline materials such as lime or sodium hydroxide.

Nutrient Requirements for Anaerobic Digestion

In order for any biological process to operrate, inorganic nutrients required by bacteria for growth must be supplied. The inorganic nutrients for growth required in the highest concentration are phosphorus and nitrogen. Since many industrial wastes are deficient in these nutrients it may be necessary to supplement the influent waste stream. The required amount of nutrients that should be added for growth can be determined from the chemical composition of the cell. Using an average cell composition of $C_5H_9O_3N$, the nitrogen requirement is about 11 percent of the cell volatile solids weight while phosphorus is about one-fifth that for nitrogen or about 2 percent of the cell weight [McCarty, 1964]. If the solids production is 1.0 gr/gr COD, the nitrogen requirement would be 11 percent of this or 0.11 gr/gr of COD while the phosphorus is 2 percent or 0.02 gr/gr of COD.

In actuality, the biological solids production and nitrogen and phosphorus requirements should be based on the fraction of waste removed during treatment rather than the waste loading [McCarty, 1964]. The reason this is not done is that in highly loaded systems the first stage of acid formation takes place to a larger extent than does the second stage of methane fermentation. Even though no waste stabilization occurs in the first stage, the acid formers still grow, and require nitrogen and phosphorus. Thus, estimates of growth and nutrient requirements based on stabilization alone may be too low.

Although nutrient requirements are low, a minimum amount of all essential nutrients such as nitrogen and phosphorus must be present. It has been observed that low COD removals resulted in an anaerobic system that did not receive the required nutrients [Lettinga and Vinken, 1980]. The minimum phosphorus:nitrogen:carbon ratio which is required for growth in anaerobic digestion treatment is approximately 1:6:100 [Jennet and Dennis, 1975].

Temperature Effects on Biological Processes

Bacteria may be divided into the following groups based on their temperature-growth ranges [O'Rourke, 1968]:

(a) Cryophiles or Psychrophiles. These may be loosely defined to include species that are able to grow rapidly below 20°C.

(b) Thermophiles. Those species growing best at temperatures above 55°C.

(c) Mesophiles. Those species with a temperature-growth range intermediate to the psychrophiles and thermophiles (20-55°C).

Most chemical and physical reactions are dependent on temperature as well as biochemical reactions catalyzed by enzymes within the above-defined biokinetic ranges. The rate of chemical reaction is a direct function of temperature, and follows the relationship developed by van't Hoff and Arrhenius [Williams, *et al*. 1978]:

$$\ln \frac{K_{T_2}}{K_{T_1}} = \frac{E(T_2 - T_1)}{R\,T_1\,T_2} \qquad (1.3)$$

where: E = activation energy, cal.

R = gas constant (1.99 cal. per °C)

T = absolute temperature, °K

K = the reaction rate constant, lit./.moles/sec.

The activation energy is that required to provide sufficient energy to a constant fraction of the molecules per unit of time in order for effective reaction to take place at given temperature [O'Rourke, 1968]. A plot of the natural logarithm of K when plotted against the reciprocal of the absolute temperature yields a straight line, with a slope of -E/R. The plot permits easy evaluation of the activation energy. It is also possible to calculate K for different temperatures provided the activation energy is known.

For chemical reactions, a temperature correction factor, Q_{10}, indicates how the reaction rate will increase if the temperature is raised by $10^{o}C$, and is defined as [Novak, 1974]:

$$Q = k, \text{Rate At } (T^{o}C + 10^{o}C) \quad (1.4)$$

It should be pointed out that the k listed in this equation is identical to the k in equation 1.3 [Sachs, et al. 1978]. Q_{10} is 2 to 3 for biochemical reactions [O'Rourke, 1968].

The curve relating growth rate to the reciprocal of absolute temperature is also linear, but only over the temperature ranger where actual growth is occurring, since the growth rate falls quickly at both the upper and lower limits of the temperature range. A sudden decrease in growth at high temperatures is often caused by thermal denaturation of proteins and possibly other cell structures such as cell membranes [Stanier, et al. 1976]. The maximum temperature for growth according to Stanier, et al. (1976) is the temperature just below which these destructive reactions become overwhelming. The lethal temperature is usually only a few degrees higher than the optimum temperature at which the growth rate is maximal.

From the effect of temperature on the rate of chemical reactions, it might be predicted that bacteria would continue to grow (although at a lesser rate) as the temperature is reduced, until the system freezes. However, Stanier, et al. (1976)

states that most bacteria cease growing at a temperature (the minimum temperature of growth) well above the freezing point of water. Each microorganism has a precise minimum temperature of growth, below which growth will not occur.

As stated by Lawrence and McCarty (1970), overall reactions carried out by the primary organisms may be quite different. (e.g., anaerobic versus aerobic conditions), but the processes of microbial growth and energy utilization are similar. The relationship between biological growth and substrate utilization has been formulated into two basic equations, where the first describes the relationship between net rate of growth of the microorganisms and the rate of substrate utilization. This equation is as follows:

$$\frac{dx}{dt} = Y\frac{dF}{dt} - bX \qquad (1.5)$$

where: $\frac{dx}{dt}$ = net growth rate of microorganisms per unit volume of reactor, mass per volume-time.

Y = growth yield cofficient, mass per mass

$\frac{dF}{dt}$ = rate of microbial substrate utilization per unit volume, mass per volume-time

= microorganisms decay coefficient, time^{-1}

X = microbial mass concentration mass per volume

The second equation [Lawrence, *et al*. 1970], which differs slightly from the form used by Monod [O'Rourke, 1968], relates the rate of substrate utilization to both the concentration of substrate and of microorganisms, and is as follows:

$$\frac{dF}{dt} = \frac{kSX}{K_s + S} \qquad (1.6)$$

where: k = maximum rate of substrate utilization per unit weight of microorganisms, time^{-1}

K_s = half velocity coefficient, equal to substrate concentration when $\frac{dF}{dt} = (1/2)\,k$, mass per volume

S = concentration of substrate surrounding the microorganisms, mass per volume

This latter equation illustrates that the functional relationship between substrates utilization and substrate concentration is continuous over the total range of substrate concentration. By combining equations 1.5 and 1.6, the following equation is developed:

$$\mu = \frac{Y\,k\,S}{K_s + S} - b \tag{1.7}$$

where: μ = $(dX/dt)/X$, the net specific growth rate, time^{-1}

There are two operational parameters which have wide application in the design and operation of biological treatment systems: U, the substrate utilization rate, and θ_c, the solids retention time. The first parameter is termed the specific utilization rate [Novak, 1974], and is represented as:

$$U = \frac{\frac{dF}{dt}\,T}{X_T} \tag{1.8}$$

where: U = specific utilization rate

$\frac{dF}{dt}\,T$ = mass of substrate utilized over a finite time

X_T = mass of organisms

The specific utilization rate is in the same general form as the Monod equation except that the Monod equation describes the microorganism's specific growth rate while equation 1.8 describes the rate of substrate uptake per unit mass. Basically, constants such as k_s (the half velocity coefficient) and k (the maximum substrate utilization rate) are dependent on temperature and substrate type. Because of this, it is necessary to determine K_s and k for each different type of substrate (acetic acid, propionic acid, etc.) and for each temperature.

Novak (1974) offers another solution, and that is to determine K_s and k for a particular substrate at a certain reference temperature and then develop a relationship that will correct equation 1.8 for temperature by inserting the proper temperature value into the equation. Novak, after reviewing investigations by O'Rourke (1968), Lawrence and McCarty (1969), Lawrence (1971), and Knowles, *et al*. (1965), found that both K_s and k vary with temperature in the following manner:

$$k_T = k_o \ell^{C_1 T} \qquad (1.9)$$

$$S = K \qquad (1.10)$$

where:

k_T = k at a reaction temperature, T

k_o = k at a reference temperature

C_1 = a constant equal to the slope of log k versus temperature line

K_{s_T} = K_s at a reaction temperature, T

K_{s_o} = K_s at a reference temperature

C_2 = constant equal to the slope of log K_s versus temperature line

This modified equation 1.8, in the following manner:

$$U = \frac{k_o \ell^{C_1 T} S}{K_{s_o} \ell^{C_2 T} + S} \qquad (1.11)$$

This equation, which describes substrate utilization, accounts for variations in temperature for either organic loading or organism concentration. When incorporated into kinetic models for various treatment processes, this equation should allow the prediction of process performance under a variety of organic loadings and operating temperatures.

The second parameter, termed sludge age, mean cell residence time or biological solids retention time, θ_c, and is defined as:

$$\theta_c = \frac{X_T}{(dX/dT)_T} \tag{1.12}$$

There are 2 θ_c values that are especially important; θ_c^M and θ_c^d. θ_c^M represents the minimum θ_c value at which complete failure of the biological process will occur. Below θ_c^M, the organisms are removed at a faster rate than they reproduce, which eventually yields total washout of the microorganisms. The value of θ_c^d represents the θ_c value to be used for design purposes and, therefore, must be greater than θ_c^M. The ratio θ_c^d/θ_c^M is defined as a safety factor for the system.

According to Lawrence and McCarty (1970), the effluent substrate concentration of a complete mix system has been found to be dependent on the solids residence time, θ_c, as shown in the following equation [Novak, 1974]:

$$\theta_c^{-1} = YU - b \tag{1.13}$$

By combining equations 1.11 and 1.13 equation 1.12 becomes [Novak, 1974]:

$$\theta_c^{-1} = \frac{Y(K_o \ell^{C_1 T})S}{K_{S_o} \ell^{C_2 T} + S} \tag{1.14}$$

When comparing U, the rate of substrate utilization by a unit mass of organisms, and θ_c, the solids residence time of the system, desired treatment efficiency can be obtained by controlling either of these two parameters. Since θ_c is the more readily measurable and easier controlled of the two, it is usually the control parameter of choice [Lawrence and McCarty, 1970].

In summary, it has been shown that chemical and physical reactions in biological processes are highly dependent on temperature. As temperature is reduced, reaction rates slow, and therefore, the rate in which the microorganisms utilize the substrate is also reduced. This means that the rate by which microorganisms utilize substrate at 35°C is higher than at 25°C. In terms of solids residence time, a higher θ_c value is required at 25°C if an indentical amount of waste stabilization is to be achieved by the microorganisms.

As mentioned earlier, the conventional single-stage digester has been used mainly to stabilize sewage sludge solids. Since, in this process, the hydraulic retention time (HRT) is equal to the solids retention time (SRT), high SRT values cannot be achieved without corresponding large digester volumes. Since the conventional process is one of the oldest and most widely used digestion process, a great deal of research has been performed on the process pertaining to temperature effects and corresponding SRT values. Following is a summary of several of these investigations, which were conducted to determine the effect of temperature on the SRT required for a desired process efficiency.

The influence of temperature on the digestion rate of sewage sludge in a batch process was originally studied by Fair and Moore in 1934 who determined, for the digestion of inoculated sludge, the necessary digestion periods at various temperatures to obtain equal solids stabilization are as follows:

at 10°C the digestion period is 90 days;

at 15°C the digestion period is 60 days;

at 20°C the digestion period is 45 days;

at 25°C the digestion period is 30 days;

at 30°C the digestion period is 20 days;

Malina (1964), using a conventional batch reactor to evaluate the effect of temperature on the mechanisms of anaerobic digestion of waste activated sludge in a completely mixed system, concluded that temperature does not affect the sequential mechanisms of digestion. However, the duration of each step and the extent to which the microbial transformations progresses is controlled by the temperature at which the digester is maintained. Digestor performance as measured by gas production showed that gas production was slightly greater at 55°C than at 35°C, when plotted against digestion time.

Rowe (1971) also performed temperature experiments with laboratory-sized conventional batch type reactors, that were fed sewage as an influent substrate. Two different reactor temperatures were used; one at 35°C and represented the mesophilic temperature zone and the other at 55°C, representative of the thermophilic temperature zone. Utilizing the volume of gas production as a measure of the process efficiency, he concluded that thermophilic digestion (at 55°C) was more rapid than mesophilic digestion. He further suggested that thermophilic digestion could be obtained with detention periods from 7 to 15 days, versus the 24 to 26 days used for mesophilic digestion.

Another temperature experiment was conducted by Stevens and Schulte (1979). Once again single-stage conventional digesters were, used with the influent waste substrate consisting of swine manure. Two basic experiments were designed to study low temperature effects on swine manure digestion. Experiment "A" was designed to compare process temperatures of 40°C and 25°C under various loading conditions, while Experiment "B" was designed to further investigate the feasibility of anaerobic digestion of swine manure at low process temperatures, In Experiment "B" each digester was kept at a constant temperature of 22.5°C and fed a constant influent substrate concentration with SRT being the only variable.

Schulte and Stevens reached two important conclusions from their experiments. The first conclusion was that the basic mechanics of the digestion process did not differ at low temperatures to those at higher temperatures. An equally important conclusion was that low temperature digestion (<25°C) was shown to require approximately twice as long an SRT to achieve the same breakdown as the normal temperature of mesophilic digestion.

There are numerous drawbacks related to the conventional single-stage digester, as described earlier in this chapter. The major drawback is that the HRT is equal to the SRT, which makes the process uneconomical for treatment of wastes containing less than 1000 mg/l BOD due to the large digester volumes required. This drawback led to the development of several new processes such as the anaerobic upflow filter, anaerobic fluidized bed and the anaerobic upflow sludge blanket process, all of which are capable of separating the SRT from the HRT, and thus making it economical to handle dilute wastes.

Not only are these new processes effectively able to separate the SRT from the HRT, but SRT values greater than 100 days are being achieved at relatively low HRT's. With the conventional single-stage digester where the HRT equals the SRT, process efficiency is controlled by the SRT. This is true because as temperatures are reduced reaction rates are also reduced, and high SRT values are required to compensate for these slower rates and, therefore, achieve comparable treatment efficiency at higher temperatures.

Since the new anaerobic treatment processes are capable of obtaining high SRT values it is understandable that they should be capable of treating organic wastes at low temperatures with high treatment efficiencies. Presently there is little information in the literature concerning this. The following is a brief review of studies conducted at low temperature using high SRT anaerobic treatment processes.

Schroepfer and Zeimke (1958) conducted tests at reduced temperatures on the anaerobic contact process, using a variety of organic wastes. They concluded that application at lower temperatures was feasible but at somewhat low loadings. Tests on several wastes at a temperature of 25°C indicated that the process was capable of removing 80 to 90 percent of the BOD load at contact unit loadings of 0.05 to 0.10 lb/ft^3/day. This was compared to results at 35°C, which showed that the unit removed 80 to 90 percent of the BOD load applied at contact unit loadings varying from 0.10 to 0.20 lb/ft^3/day.

Dague, et al. (1966), concluded that there was no additional difficulty in operating the contact process at 25°C. They found gas production decreased with the same degree of waste stabilization, indicating a slower rate of endogenous metabolism of the microorganisms. The higher solids concentration tended to confirm this observation.

In 1979, a pilot-sized upflow anaerobic sludge blanket process (UASB) was intalled in the Netherlands to treat liquid sugar waste [Pette, et al. 1979]. During the first three months of the year the plant was fed irregularly, due to troubles with the heating-system and cold weather. For this reason, the digester contents were kept at relatively low temperatures (between 25°C and 30°C). Even under these conditions, it was possible to treat a COD concentration of about 10,000 mg/l at a loading of approximately 10 kg COD/m^3/day, and an efficiency of 96 percent.

Additional low temperature studies using the UASB process were conducted by Lettinga and Vinken (1980) in the years previous to 1980. During the study several different wastes were treated; among them a dried potato waste as well as a volatile fatty acid waste. From their results, these workers were able to conclude that there was a sigificant dependency on temperature of the maximum specific activity (gr COD/gr VSS/day) attained, no matter which waste was being treated. The maximum specific activity represents the total amount of waste (gr, COD) that a unit amount of active biomass (gr. VSS) can stabilize in a 1 day period. Their results showed that as the temperature in the reaction vessel was lowered, the value of the maximum specific activity also decreased. This phenomena was attributed to a decrease in the biological rates of the microorganisms involved.

Finally, Jewell and Switzenbaum (1978) conducted low temperature studies using an anaerobic attached film expanded bed reactor (AAFEB) for synthetic waste with glucose and yeast as the main carbon source. In their study, temperatures of 10 °C, 20 °C and 30 °C were used. For each temperature, 6 hydraulic retention times ranging from 6 hours to 20 minutes were evaluated. At each of the 6 hydraulic retention times three different influent substrate concentrations (200, 400 and 600 mg/l COD) were used, providing 3 different organic loading rates for each hydraulic extension time.

Jewell and Switzenbaum concluded that the efficiency of substrate utilization was a function of the hydraulic retention time and organic loading rate, as well as influent substrate concentration and temperature. They also concluded that the influence of temperature on process efficiency is not as pronounced as one might expect, since the systems showed good compensation by developing increased biomass concentrations at lower temperatures.

From the above studies, it would appear that temperature effects are not as severe on high SRT systems as they are on lower SRT, conventional digestion, systems. The reason for this phenomenon can be attributed to the fact that slower retention rates are compensated by the high SRT of the system, which is similar to the effects of toxic compounds described in Chapter 3. Dague, _et al_. (1966) suggested that the reason for higher microorganism concentration at low temperature results from a reduced rate of endogenous metabolism by the microorganism involved. The general trend seems to point to the fact that low temperature, high SRT systems are able to perform on an equal to possibly better basis than the high temperature, high SRT system, up to maximum volumetric loading of the system at a given temperature. The maximum value represents the volumetric loading value where the higher solids retention time of the low temperature reactor will no longer be able to offset the lower reaction rates at the lower than optimum temperatures which results in lower process efficiencies.

REFERENCES

Bhaskaran, T.R. and Chakrabarty, R.N. 1966. Pilot Plant for Treatment of Cane-Sugar Waste. _J. Wat. Poll. Con. Fed_. 38:1160-1169.

Buswell, A.M. and Mueller, H.F. 1952. "Mechanisms of Methane Fermentation". _Ind. Engr. Chem._ 44:550-552.

Chow, V.T., Eliassen, R. and Linsley, R. Eds. 1972. _Wastewater Engineering: Collection, Treatment and Disposal_. McGraw-Hill Book Co., New York City.

Dague, R.R., McKinney, R.E. and Pfeffer, J.T. 1966. "Anaerobic Activated Sludge." _J. Wat. Poll. Con. Fed_. 38:220-226.

DeWalle, F.B. and Chian, E.S.K. 1976. "Kinetics of Substrate Removal in a Completely Mixed Anaerobic Filter." _Biotech. Bioeng_. 18:1275-1295.

Fair, G.M. and Moore, E.W. 1934. "Time and Rate of Sludge Digestion and Their Variation with Temperature." _Sew. Works J_. 6:3-12.

Gorgan, J.M. and Obayashi, A.W. 1980. Unpublished Data, Illinois Institute of Technology, Chicago, Ill.

Heertjes, P.M. and van der Meer, R.R. 1978. "Dynamics of Liquid Flow in an Upflow Reactor Used for Anaerobic Treatment of Wastewater." _Biotech. Bioeng_. 20:1577-1594.

Jennet, J.C. and Dennis, N.D. 1975. Anaerobic "Filter Treatment of Pharmaceutical Waste." _J. Wat. Poll. Con. Fed_. 47:104-121.

Jeris, J.S., Owens, R.W., Hickey, R. and Flood, F. 1977. "Biological Fluidized Bed Treatment for BOD and Nitrogen Removals." _J. Wat. Poll. Con. Fed_. 49:816-831.

Jewell, W.J., 1980. "Development of the Attached Microbial Film Expanded Bed Process for Aerobic and Anaerobic Waste Treatment." Cornell University, Ithaca, New York.

Jewell, W.J., Switzenbaum, M.S. and Morris, J.W. 1981. "Municipal Wastewater Treatment with the Anaerobic Attached Microbial Film Expanded Bed Process." _J. Wat. Poll. Con. Fed_. 53:482-490.

Joseph Oat Corporation. 1981. "Heileman Wastewater Tests Spell Success." _Biothane Digest_. 1:1-4.

Joseph Oat Corporation. 1981. "Ore-Ida Food Tests Are In." _Biothane Digest_. 1:1-4.

Knowles, G., Downing, A.L. and Bartlett, M.J. 1965. "Determination of Kinetic Constants for Nitrifying Bacteria in Mixed Cultures with the Aid of an Electric Computer." _J. Gen. Microbiol_. 38:263-278.

Lawrence, A.W. 1971. "Application of Process Kinetics to Design of Anaerobic Processes." Advanced Chemical Series No. 195.

Lawrence, A.W. and McCarty, P.L. 1969. "Kinetics of Methane Fermentation in Anaerobic Treatment." _J. Wat. Poll. Con. Fed_. 41:R1-R16.

Lettinga, G. 1979. "Direct Anaerobic Treatment Handles Waste Effectively." _Ind. Wastes_. 53:18-24.

Lettinga, G. and de Zeeuw, W. 1980. "Acclimation of Digested Sewage Sludge During Start-Up of an Upflow Anaerobic Sludge Blanket (UASB) Reactor." Paper presented at the 35th Purdue Industrial Waste Conference, May 16-18, 1980, at Purdue University, West Lafayette, Indiana.

Lettinga, G., van der Geest, A. Th., Hobma, S. and Laan, J. 1979. "Anaerobic Treatment of Methanolic Wastes." _Wat. Res_. 13:725-737.

Lettinga, G., Velsen, A.F., de Zeeuw, W. and Hobma, S.W., 1979. "Feasibility of the Upflow Anaerobic Sludge Blanket (UASB) Process." _Enve. Engr_. 1979:35-45.

Lettinga, G., Velsen, A.F., Hobma, S.W., de Zeeuw, W. and Klapwijk, A. 1980. "Use of the Upflow Sludge Blanket, (UASB) Reactor Concept for Biological Wastewater Treatment." _Biotech. Bioeng_. 22:699-734.

Lettinga, G. and Vinken, J.N. 1980. "Feasibility of the Upflow Anaerobic Sludge Blanket (UASB) Process for the Treatment of Low Strength Wastes." Paper presented at the 35th Purdue Industrial Waste Conference, May 16-18, 1980, at Purdue University, West Lafayette, Indiana.

Malina, J.F. 1964. "Thermal Effects on Completely Mixed Anaerobic Digestion." Sew. Works J. 28:52-56.

McCarty, P.L. 1964. "Anaerobic Waste Treatment Fundamentals, Part I." Pub. Works 95 (Sept.):107-112.

McCarty, P.L. 1964. "Anaerobic Waste Treatment Fundamentals, Part II." Pub. Works 95 (Oct.):123-126.

McInerney, M.J., Bryant, M.P. and Stafford, D.A. 1979. "Metabolic Stages and Energetics of Microbial Anaerobic Digestion." Paper presented at the First International Symposium on Anaerobic Digestion, September, 1979, University College, Cardiff. Wales, Great Britain.

McNary, R.R., Wolford, R.W. and Dougherty, M.H. 1962. "Experimental Treatment of Citrus Wastewater." Paper Presented at the 8th Purdue Industrial Waste Conference, May, 1962, at Purdue University, West Lafayette, Indiana.

Novak, J.T. 1974. "Temperature-Substrate Interactions in Biological Treatment." J. Wat. Poll. Con. Fed. 46:1984-1994.

Obayashi, A.W. and Roshanravan, M. 1980. Unpublished Data, Illinois Institute of Technology, Chicago Ill.

Obayashi, A.W., Stensel, H.D. and Kominek, E. 1980. "Anaerobic Treatment of High Strength Industrial Wastewaters. Paper presented at the 1980 American Institute of Chemical Engineering Meeting, May, 1980, Chicago, Ill.

O'Rourke, J.T. 1968. "Kinetics of Anaerobic Waste Treatment at Reduced Temperatures." Ph.D. Thesis, Stanford University, Palo Alto, California.

Pettet, A.E., Tomlinson, T.G. and Hemens, J. 1959. "The Treatment of Strong Organic Wastes by Anaerobic Digestion." Paper presented at a Meeting of the Institution of Public Health Engineers, April 27, 1959, at the Conference Room of Hotel Majestic, New York, N.Y.

Pette, K.C. 1979. "Anaerobic Waste Water Treatment at CSMN-Sugar Factories." Paper presented at the 16th General Assembly-Commission Internationale Technique de Sucrerie, 1979, Amsterdam, Netherlands. In Reprint.

Pette, K.C., Heertjes, P.M., de Vletter, R. and van der Meer, R.R. 1979. "The Upflow Reactor for Anaerobic Treatment of Wastewater Containing Fatty Acids." CSM Suiker, Amsterdam, Netherlands.

Pette, K.C., de Vletter, R., Wind, E. and van Gils, W. 1979. "Full-Scale Anaerobic Treatment of Beet-Sugar Wastewater. CSM Suiker, Amsterdam, Netherlands.

Pette, K.C., Sax, R.J. and Holtz, M. 1980. "Production of Biogas from Wastewaters of Food Processing Industries." Paper presented at the Conference on Industrial Energy Conservation Technology and Exhibition, April 13-15, 1980, Houston, Texas.

Pfeffer, J.T. 1979. " Anaerobic Digestion Processes." Paper presented at the First International Symposium on Anaerobic Digestion, September, 1979 at University College, Cardiff, Wales, Great Britain.

Ross, W.R. 1980. "Treatment of Concentrated Organic Industrial Wastes by Means of the Anaerobic Digestion Process: A Review of South African Experience and Current Research." In Xerox. Die Institut Vir Dir Bestryding Van Waterbesoedeling.

Rowe, D.R. 1971. "Anaerobic Sludge Digestion, Mesophilic-Thermophilic." Sew. Works J. 35:74-76.

Sachs, E.F., Jennet, J.C. and Rand, M.C. 1978. "Anaerobic Treatment of Synthesized Organic Chemical Pharmaceutical Wastes." Paper presented at the 33rd Purdue Industrial Waste Conference, May 9-11, 1978, at Purdue University, West Lafayette, Indiana.

Schroepfer, G.F., Fullen, W.J., Johnson, A.S., Ziemke, N.R. and Anderson, J.J. 1955. "The Anaerobic Contact Process as Applied to Packing House Waste." Sew. Ind. Wastes. 27:460-486.

Schroepfer, G.J. and Ziemke, N.R. 1958. "Development of the Anaerobic Contact Process." Sew. Ind. Wastes. 31:697-711.

Stander, G.J. 1967. "Treatment of Wine Distillery Wastes by Anaerobic Digestion." Paper presented at the 22nd Purdue Industrial Waste Conference, May, 1967, at Purdue University, West Lafayette, Indiana.

Steffen, A.J. and Bedker, M. 1961. "Operation of Full-Scale Anaerobic Contact Treatment Plant for Meat Packing Wastes." Paper presented at the 16th Purdue Industrial Waste Conference, May, 1961, at Purdue University, West Lafayette, Indiana.

Stevens, M.A. and Schulte, D.D. 1979. " Low Temperature Anaerobic Digestion of Swine Manure." J. Amer. Soc. Civ. Engr. San. Div. 105:33-42.

Stobaugh, R. and Yergin, D. Eds. 1979. Energy Future. Random House, Inc., New York, N.Y.

Switzenbaum, M.S. and Jewell, W.J. 1978. "Anaerobic Attached Film Expanded Bed Reactor Treatment of Dilute Organics." Paper presented at the 51st Annual Water Pollution Control Federation Conference, October 2, 1978, Anaheim, California.

Tait, S.J. and Friedman, A.A. 1981. "Anaerobic Rotating Biological Contactor for Carbonaceous Wastewaters." J. Wat. Poll. Con. Fed. 52:2257-2269.

Williams, V.R., Mattice, W.L. and Williams, H.G. 1978. _Basic Physical Chemistry for the Life Sciences_. W.H. Freeman and Company, San Francisco, California.

Witt, E.R., Humphrey, W.J. and Roberts, T.F. 1979. "Full-Scale Anaerobic Filter Treats High Strength Wastes. " Paper presented at the 34th Purdue Industrial Waste Conference, May 8-10, 1979, at Purdue University, West Lafayette, Indiana.

Young, J.C. and McCarty, P.L. 1968. "The Anaerobic Filter for Waste Treatment." Technical Report NO. 87, Dept. of Civil Engineering, Stanford University, Palo Alto, California.

CHAPTER 2

BIODEGRADATION OF INDUSTRIAL ORGANIC COMPOUNDS BY ANAEROBIC PROCESSES

INTRODUCTION

Recent advances in chemical technology have led to the production of many and new potentially dangerous compounds. Some, through normal usage, eventually end up as a constituent in wastewater. These new substances, representing a wide array of compounds ranging from phenols to pesticides, each presents its own individual problems when dealing with its ultimate disposal. This has led to a great many problems and challenges for the people who must deal with these wastes, namely the wastewater treatment industry. They have responded by introducing many new and innovative techniques for the treatment of these wastes, many of which employ anaerobic processes. As in the case of aerobic processes, not all organic compounds will satisfactorily undergo breakdown through anaerobic conditions.

DEFINITION OF BIODEGRADATION

The process of biodegradation is not an inherent characteristic of a specific organic compound. Rather, it is the result of a highly complex chain of events revolving around interactions between the specific compound, microorganism, and the environmental conditions where the two are combined [Kirsch, *et al*. 1981]. Due to the highly variable nature of the process and the broad scope and boundaries that must be included under the term biodegradation, a precise definition is difficult to formulate. Fundamentally, one might consider biodegradation as a change in the parent compound's chemical or physical properties, brought about by biochemical processes [Standard Methods Committee, 1967].

This change usually involves conversion of a specific chemical compound or group of compounds, and occurs by a series of discreet steps. It is accompanied by a corresponding decrease in the energy of the parent compound [Standard Methods Committee, 1967]. The degree of change can be divided into the following three operational categories [Barth and Bunch, 1979]:

1. Primary Biodegradation: The minimal amount of change necessary in the chemical composition of the parent compound to make it no longer detectable by the analytical measurement technique used in the detection of the parent compound.

2. Ultimate Biodegradation: The mose desirable endpoint in any biodegradation reaction, involving the total conversion of the parent compound by anaerobic processes into CO_2, H_2O, and inorganic substances (if present in the parent compound). If broken down anaerobically, the end products would include; CH_4 and H_2O in addition to CO_2 and inorganic residuals.

3. Acceptable Biodegradation: In this category are placed biodegradation reactions where the parent compound is changed to the minimum extent so that some desirable property such as toxicity is no longer present.

The degree of stabilization strived for depends on the compound and its concentration. In some cases, primary degradation may be perfectly acceptable for receiving waters while in other cases anything short of ultimate degradation would be undesirable.

TESTS USED TO PREDICT BIODEGRADATION

A wide variety of tests are available for measuring biodegradability. Recent interest in biodegradability studies, especially by the soap manufacturing industry, has led to new test procedures. All present existing tests can be divided into two categories; those to assess biodegradability potential, and those that simulate conditions found in wastewater treatment processes. Biodegradability potential tests (or presumptive tests) are designed to be used as quick screening procedures to assess the susceptability of the test compound to biological breakdown. These tests are simple to run, and equipment needs are minimal. Simulation tests on the other hand are more difficult to perform, but provide more realistic conditions. If results of a potential test are inconclusive, a simulation test should be performed.

Regardless of the test types, it must be emphasized that the nature of the test conditions will influence the test outcome to a degree [Gilbert and Lee, 1980]. This necessitates proper control over environmental and nutritional facets of the test. Besides ensuring the promulgation of a healthy population of microorganisms, other factors such as pH and temperature should be carefully controlled.

In testing for biodegradability one should keep in mind two additional processes; acclimation and co-metabolism. Both processes can play a major role in determining the test outcome. Acclimation, the process by which bacteria adapt their metabolic activities to degrade foreign substances, occurs by mutation and natural selection or by induction of specific enzymes. The degree of acclimation is highly dependent on the number and types of microorganisms brought into contact with the test compound. The second process, co-metabolism is equally important. Some substances will not be degraded unless some additional factor is present. Such is the case if the ABS branched chain tetrapropylene benzene sulphonate and an unidentified species of _Pseudomonas_. In a pure culture the _Pseudomonas_ will not actively degrade the surfactant. However, upon the addition of glucose the surfactant will be degraded to isoproponal and catechol [Pointer, 1974]. Testing systems involving co-metabolism are usually characteristic of environmental conditions encountered in nature and, therefore, provide more realistic estimates of biodegradation [Barth and Bunch, 1979].

In the next section, a simple review is presented of the most widely used tests to determine whether or not a compound is biodegradable. Most of these tests are presented in their aerobic form, but all with the exception of the activated sludge test can be easily run under anaerobic conditions, with certain modifications.

The River Die Away Test

The river die away test is a widely used screening test to determine biodegradability potential. It is based on the principal assumption that by exposing the test compound to a mixed microbial population taken from a natural source, conditions encountered in nature can be realistically simulated [Payne, et al. 1970]. The test itself is simple to perform. Dilute solutions of the test material are added to samples of river water. These samples are aerated and allowed to stand under quiescent conditions at room temperature. Periodically, samples are withdrawn and chemically analyzed to monitor the percentage of test compound remaining. The resulting data is used to plot a die away, or disappearance, curve. Besides determining if a product is biodegradable, this procedure also establishes a rate at which the product is consumed. However, it must be understood that this rate is only relative. If more of the test compound is added to the sample container, the preselected population of microorganisms could degrade the additional test material at a faster rate [Barth and Bunch, 1979]. The test can take long periods of time (up to 30 days), but this can be overcome by adding a small amount of activated sludge to speed up reaction rates [Sweeny and Foote, 1964]. A similar test can be run anaerobically. In the anaerobic version, sanitary sewage devoid of any O_2 is substituted for river water [Vath, 1964].

The advantages and disadvantages of this test are as follows [Barth and Bunch, 1979; Bunch and Chambers, 1967; Weaver and Coughling, 1964]:

Advantages

1. Gives the relative rate as well as the completeness of the disappearance of the test compound.

2. Requires little equipment and is simple to run.

3. Large numbers of compounds can be screened in a relatively short time.

Disadvantages

1. Every river will not necessarily yield the same results. Factors such as pH, nutrient content, and the number and types of microorganisms present in the water at the time of testing will significantly alter results.

2. It may be necessary to analyze for specific degradation products, since the analytical procedure employed may not detect slight modifications of the parent compound to products that are resistant to further degradation.

3. Seasonal effects will substantially alter results from the same river.

The Warburg Respirometer

For measuring the oxygen consumed in the aerobic microbial degradation of organic compounds, the Warburg Respirometer is a readily available and widely used device. It can also be adopted to measure gases evolved from the breakdown of organic compounds by anaerobic processes [Kugelman and McCarty, 1965; McCarty, *et al*. 1963; Gossett and McCarty, 1976]. The Warburg apparatus is extremely useful when no specific analytical technique for the measurement of the test compound exists [Barth and Bunch, 1979]. Using the Warburg apparatus, the test for biodegradability is as follows: The test compound along with a suitable acclimated seed is placed in a test flask. A manometer is attached at the outlet of the flask. At a constant temperature and volume, the amount of gas consumed (or produced) is represented by changes in pressure. By measuring the pressure change, and knowing the gas and fluid volumes of the flask, the temperature, the gas being exchanged and the density of the fluid in the manometer, the amount of gas evolved or consumed can be calculated [Umbreit, *et al*. 1963].

The advantages and disadvantages of this method are as follows [Barth and Bunch, 1979; Owen, *et al*; McCarty, *et al*. 1963].

<u>Advantages</u>

1. The apparatus has the capability to run many samples simultaneously.

2. Changes in rate and lag periods can be measured.

<u>Disadvantages</u>

1. The Warburg unit is costly and requires trained personnel to operate.

2. Small sample sizes make it difficult to obtain representative samples of the wastewater, which makes subsequent chemical analysis difficult.

3. It is difficult to collect samples from the gas or liquid phases during the test.

Flask Test Method

Second only in popularity to the activated sludge test (discussed later), the flask test method is a widely applicable mtehod for the assessment of biodegradability. The simplicity of the test and the fact that its equipment needs are minimal make this an attractive presumptive test. Many versions of this test are in use today [Bunch and Chambers, 1967; SDA, 1965; Bunch and Barth, 1979]. The basic test is outlined by the Soap and Detergent Association for use in presumptive testing is as follows [SDA, 1965]. An inocula of microorganisms is placed in a flask along with a suitable amount of chemical medium composed of a basal media and the test compound. After two-72 hour adaptive transfers to fresh media, an inocula is transferred to fresh media again and the reduction of the test compound is observed. The test inoculum can be obtained from many sources including natural sewage, lab cultures, or commercially obtainable cultures [SDA, 1965]. The flask can be shaken to ensure adequate gas exchange [SDA, 1965], or allowed to remain static [Bunch and Chambers, 1967]. In all cases a control of a known biodegradable substance should be run alongside this test, to ensure that suitable conditions for biodegradtion to proceed are maintained.

Modifications of this test exist for use in anaerobic situations. The biochemical methane potential test is one such modification [Owen, *et al*. 1979]. In this test sample biodegradability is determined by measuring the cumulative methane production from an incubated sample. The results can be related to sample volume (m^3CH_4/m^3) and sample mass (m^3CH_4/kg COD). The percentage of organic matter converted to methane can be determined by using an empirical factor of 0.35 m^3CH_4 (at STP) produced per kilogram of COD removed [McCarty, 1964].

This test offers the following advantages and disadvantages [Bunch and Chambers, 1967; USDA, 1965; Barth and Bunch, 1979]:

Advantages

1. The test is simple to run and requires little equipment.

2. Reproducibility of results are fairly good.

3. Its simplicity makes it deal as a presumptive test.

Disadvantages

1. Microorganisms must have had prior exposure to test compound of an acclimation period must be incorporated into the test.

2. A similar reference standard needs to be run as a control.

3. Specific analytical techniques must be available to measure for suspected partial degradation products.

Activated Sludge

Among all the available biodegradation tests that exist today, the activated sludge test most closely simulates the conditions encountered in most sewage treatment plants. The activated sludge test is basically a laboratory scale model of the larger activated sludge units found in many conventional treatment plants. Much of the work done in the past with design and operation of activated sludge systems is applicable to laboratory size models. Testing of compounds involves placing the test compound, a suitable inoculum, and an appropriate feed in a reactor vessel. The vessel is then subjected to a daily aeration period of which the time span is variable but, in most cases, lasts for 23 hours with settling being allowed to occur for 1 hour [SDA, 1965].

Many variations on the above theme of operation are in use today. These variations include the semi-continuous, continuous, German, and porous pot models. In the semi-continuous technique, after a 23 hour period of aeration and 1 hour period of settling the supernatant is withdrawn, and fresh feed with a given concentration of test compound is added to bring the vessel up to the starting volume. This sequence of events continues daily until test completion. The confirmatory test used by the SDA is of this mode [SDA, 1965]. The continuous technique, as the name implies, runs constantly and is continuously fed additional substrate as the supernatant is drawn off [Sweeny and Foote, 1964]. The German test is a modified continuous feed system. The interesting variation is that no seed material is added at the beginning of the test. The seed must develop from airborne bacteria [Weaver and Coughlin, 1964]. The porous pot technique, introduced by Hopwood and Downing (1965), modifies the German test [Stennett and Eden, 1971]. The test setup is exactly the same, except that the official German aeration vessel is replaced by a porous polythene vessel. This modification retains the suspended flocs, eliminating the need for a settling apparatus [Stennett and Eden, 1974]. This makes the apparatus more amenable to submersion in water baths for temperature control.

The advantages and disadvantages of the activated sludge test are as follows [Barth and Bunch, 1979; Sweeny and Foote, 1964; Stennett and Eden, 1971; Weaver and Coughlin, 1965]:

<u>Advantages</u>

1. This method closely simulates actual wastewater treatment characteristics.

2. An unlimited sample size can be used.

3. The equipment needs are slight and most equipment should be found in well stocked laboratories.

4. The ability to reproduce the results is fairly good.

Disadvantages

1. Test compound can adsorb to the activated sludge.

2. Complete acclimation of sludge to feed can be a slow and time consuming process.

3. Specific chemical detection methods are required to identify compounds present in wastewater feed.

BIODEGRADATION OF ORGANIC COMPOUNDS UNDER ANAEROBIC CONDITIONS

The following section is a brief summary of research conducted in the area of biodegradation of organic compounds under anaerobic conditions. It is not intended to be a detailed description of all past work, but rather to be used as a tool for the development of further applications of anaerobic systems in treatment of industrial organic waste. Important areas of research are identified and proposed biological pathways are presented when available. This section places major emphasis on treatment of the EPA priority pollutants (see Table 2.1), but other wastes are also discussed. The range of compounds covered is divided into the following categories:

1. Aromatic compounds

2. Pesticides

3. Agricultural waste

4. Soap and detergents

5. Miscellaneous priority pollutants

6. Miscellaneous organic compounds

TABLE 2.1

PRIORITY POLLUTANTS
Established by the EPA
(after Hutton, 1980)

CHLORINATED ALKANES
Methyl Chloride
Methylene Chloride
Methyl Bromide
Chloroform
Bromoform
Carbon Tetrachloride
Dichlorobromomethane
Trichlorofluoromethane
Dichlorodifluromethane
Chlorodibromomethane
Chloroethane
1,1-Dichloroethane
1,2-Dichloroethane
1,1,1-Trichloroethane
1,1,2-Trichloroethane
1,1,2,2-Tetrachloroethane
Hexachloroethane
1,1-Dichloroethylene
1,2-trans-Dichloroethylene
1,2-Dichloropropane
1,2-Dichloropropylene
Trichloroethylene
Tetrachloroethylene
Vinyl Chloride
Hexacholorobutadlene
Hexachlorocyclopentadlene

CHLORINATED AROMATICS
1,2,4-Tricholorobenzene
Chlorobenzene
Hexachlorobenzene
2-Chloronaphthlanene
1,2-Dichlorobenzene
1,3-Dichlorobenzene
1,4-Dichlorobenzene

CHLORINATED ESTHERS
bis(Chloromethyl)Esther
2-Chloroethyl Vinyl Esther
4-Bromophenyl Phenyl Esther
bis(2-Chloroethox)Methane
bis(2-Chloroethyl)Esther
4-Chlorophenyl Phenyl Esther
bis(2-Chlorolsopropyl)Esther

AROMATICS
Benzene
Toluene
Ethylbenzene
Naphthalene
Fluoranthene
Acenaphthene
Benzo (a) Anthracene
Benzo (a) Pyrene
Chrysene
Indeno(1,2,3-c,d)pyrene
3,4-Benzofluoranthene
Benzo(K)fluoranthene
Acenaphthylene
Benzo(ghl)perylene
Fluorene
Phenanthene
Dibenzo(a,h)anthracene
Pyrene

PHTHALATE ESTERS
bis(2-ethylhexyl)phthalate
butyl benzyl Phthalate
Di-n-butyl Phthalate
Di-n-octyl Phthalate
Diethyl Phthalate
Dimethyl Phthalate

PHENOLS
Phenol
2-Chlorophenol
2,4-Dichlorophenol
Pentachlorophenol
2-Nitrophenol
2,4-Dimethylphenol
4-Nitrophenol
2,4-Dinitrophenol
4,6-Dinitro-o-cresol
2,4,6-Trichlorophenol
para-Chloro-meta-Cresol

SUBSTITUTED AROMATICS
Nitrobenzene
2,4-Dinotrotoluene
2,6-Dinitrotoluene
2,3,7,8-Tetrachlorodi-
benzo-p-Dioxin
Benzidine
3,3-Dichlorobenzidene
1,2-Diphenyl Hydrazine

POLYCHLORINATED BIPHENYLS
PCB-1242 PCB-1221 PCB-1260
PCB-1254 PCB-1232 PCB-1016

MISCELLANEOUS
Acrolein
Acryloritrile
Asbestos
Cyanide
Isophorone
N-Nitrosodimethylamine
N-Nitrosodiphenylamine
N-Nitrosodi-n-propylamine

PESTICIDES
Aldrin
Diedrin
Chlordane
4,4-DDT
4,4-DDE
4,4-DDD
a-Endosulfan-alpha
b-Endosulfan-beta
Endosulfan Sulfate
Endrin
Endrin Aldehyde
Heptachlor
Heptachlor Epoxide
e-BHC - Alpha
b-BHC - Beta
r-BHC(Lindane)-Gamma
g-BHC - Delta
Toxaphene

METALS
Antimony
Arsenic
Beryllium
Cadmium
Chromium
Copper
Lead
Mercury
Nickel
Selenium
Thallium
Zinc

Aromatic Compounds

The aromatic compounds comprise a wide variety of substances characterized by the presence of at least one benzene ring. Members of this group include phenols, chlorinated aromatics, and substituted aromatics. 41 of the 129 priority compounds listed in Table 2.1 are from this group. Many aromatic substances are toxic in small amounts, and many are carcinogenic. The use of aromatic compounds is widespread in industry and they appear as constituents in wastewater originating from a number of industrial sources including coal gasification plants, oil refining, plastics, petrochemicals, pharmaceuticals and paper mills. Many of these industries emit waste streams which are comprised of a mixture of aromatics, some of which have been found to be inhibitory to aerobic biological reactions [Neufeld and Mack, 1980]. This aspect makes anaerobic processes an attractive area to explore. It is known that anaerobic breakdown of the benzene nucleus can occur by two different pathways; they are: 1) photometabolism and 2) methanogenic fermentation. Each process will be discussed individually.

Anaerobic Photometabolism of Aromatic Compounds

Procter and Scher (1980) first to reported the isolation of a group of photosynthetic bacteria of the genus Rhodopseudomonas which were capable of metabolizing benzoate anaerobically. Using both resting cell suspensions and cell extracts, they analyzed for metabolic intermediates and from their data proposed the following pathway, with the implication that molecular oxygen could be used to replace the photochemically generated oxident:

benzoate→protocatechuate→catechol→ α-oxo acid

Dutton and Evans (1967) challenged some conclusions of Procter and Scher, based on research conducted with cultures of Rhodopseudomonas palustris. It was shown that no significant oxygen uptake occurred when R. palustris was grown on benzoate, and attempts to grow R. palustris under aerobic conditions with benzoate as a substrate resulted in failure. In addition, it was found that the direct addition of air into an actively photometabolizing benzoate culture in the cessation of benzoate utilization.

Further studies by Dutton and Evans (1969) resolved the question of how the anaerobic metabolism of benzoate occurred. Since the reaction was anaerobic, the use of an oxygenative ring cleavage was not plausible. The only other feasible biochemical options available for ring fissure would be hydrogenation or hydration. Proceeding under the assumption that hydrogenation was responsible, expected intermediates were incubated with cells actively photometabolizing labelled (O ^{14}C) benzoate. At intervals samples were withdrawn, and analyzed for possible isotopic exchanges. The results of experiments of this type led to the identification of several compounds suspected of being intermediates. A pathway was proposed, which consisted of a reduction of benzoate (and derivatives) to cyclohexane-carboxylate derivatives, followed by a coenzyme A mediated beta-oxidation. A pathway of this type received independent support based on findings of Guyer and Hegeman (1969). The proposed pathway is pictured in Figure 2.1. Benzoate (I) is reduced to cyclohexanecarboxylate (II) by the addition of H_2, which in turn is dehydrogenated by the removal of H_2 to give cyclohexylene-carboxylate (III), the hydration of which results in the formation of 2-hydroxycyclohexane-carboxylate (IV) which is in turn dehydrogenated to oxocyclohexanecarboxylate (V). This, by hydration, results in ring cleavage forming the compound pimelate (VI). It was posed that the reaction is catalyzed by reductases coupled to the redox potential component of the light induced electron transport systems, and it was further suggested that the compound ferredoxin served this function. However, Whittle (1975) showed that ferredoxin did not serve this role. While search goes on for a possible catalyst, it is believed that all initial degradation of aromatic compounds under anaerobic conditions proceed by the above mechanism.

Methanogenic Fermentation of Aromatic Compounds

Early evidence that biological destruction of benzoid structures in strictly anaerobic conditions was possible was provided by Tarvin and Buswell (1934). Until this time, no evidence existed that the ring structure could be broken by anything other than an oxygenative ring cleavage. Using a sewage sludge inoculum, it was shown that benzoate, phenylacetate, phenylpropionate, and cinnamate were

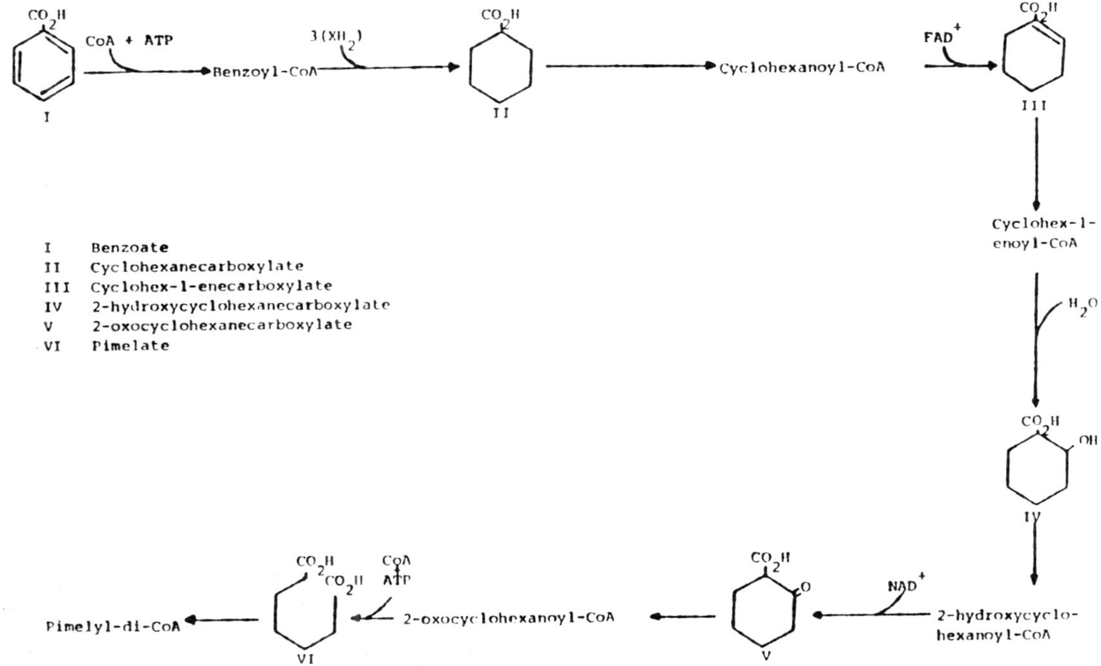

I Benzoate
II Cyclohexanecarboxylate
III Cyclohex-1-enecarboxylate
IV 2-hydroxycyclohexanecarboxylate
V 2-oxocyclohexanecarboxylate
VI Pimelate

Figure 2.1 Proposed Pathway for the Reduction of Benzoate (Whittle and Evans 1975) (Dutton and Evans 1969).

completely degraded to CO_2 and CH_4. Lower fatty acids were detected as reaction intermediates.

The work of Clark and Fina (1952) confirmed these results. Using a sewage sludge inoculum, they reported that a methanogenic culture of bacteria could degrade benzoic acid to CH_2 and CO_4, when benzoic acid was used as both a carbon and energy source. The amount of gas production was as high as 90 percent of the theoretical yield. Extremely long acclimation times, some as long as 5 weeks, were required before gas production was intiated. This time period was found to be shortened by adapting the bacteria to an acetic acid substrate before adapting them to benzoic acid. Gas analysis showed that hydrogen gas was not present in appreciable amounts, indicating that methane is formed by some mechanism other than CO_2 reduction. In an attempt to determine if either catechol or protocatechnic acid functioned as an intermediate, both were fed to stabilized benzoate-utilizing cultures. Cessation of gas production suggested that the pathway for anaerobic degradation of benzoic acid is different than the aerobic pathway proposed by Stanier (1948, 1950a, 1950b).

Conclusive evidence of the anaerobic degradation of benzoic structures to CH_4 and CO_2 was supplied by Fina and Fiskin (1960). Using C^{14} labelled benzoic acid as a substrate, labelled CH_4 and CO_2 were detected as the end products of the reaction. Specific label attachment showed that the carboxyl carbon of benzoate behaves similar to exogenous CO_2, in that it is not primarily reduced to methane. The number one carbon of benzoic acid was found to be converted primarily to methane.

Further experiments with labelled benzoic acid by Nottingham and Hungate (1960) confirmed the results of Fina and Fiskin. Using radioisotopes, they found that the composition of the gases evolved under anaerobic conditions agreed with the following balanced equation:

$$4\ {}^{14}C_6H_6CO_2 + 18\ H_2 \rightarrow 15\ {}^{14}CH_4 + 9\ {}^{14}CO_2 + 4CO_2 \quad (2.1)$$

Chmielowski, <u>et al</u>. (1965a, 1965b, 1966) conducted experiments on the decomposition of phenol, p-cresol and resorcinol. Using batch reactors with 100 mg/l of the respective phenolic compound, mixed populations of bacteria yielded complete conversions to CH_4 and CO_2. Analysis of the accumulated intermediates revealed them to be saturated compounds, and not volatile acids as one would expect. It was therefore postulated that degradation of phenols proceeded first by hydrogenation of the aromatic structure to form alicyclic intermediates. These intermediates underwent ring fission, followed by the methanogenesis of the degradation products.

Keith (1972, 1978) using isotopic trapping experiments identified cyclohexanecarboxylic acid, 1-cyclohexene-1-carboxylic acid, heptanoate, valerate, butyrate, propionate, and acetate as reaction intermediates in the degradation of benzoic acid. Based on these results, he proposed the pathway shown in Figure 2.2 Benzoate is reduced in a stepwise progression to 1-cyclohexene-1-carboxylic acid which undergoes fission of the cyclohexenecarboxylic ring by the addition of two hydrogen atoms to form heptanoic acid. Heptanoic acid is further catabolized by beta oxidation to valerate, which is reduced to propionate with the simultaneous formation of acetic acid.

Studies by Ferry and Wolfe (1976) enhanced the understanding on the degradation of benzoate. From actively degrading benzoate cultures they isolated three types of bacteria; *Methanobacterium formicum*, *Methanospirillum bungatii*, and an unidentified acetate degrader. *M. formicum* and *M. bungatii* were tested to determine if either could actively degrade benzoate in pure culture. The negative results confirm Chmielowski's findings, and suggest that the function of the methanogenic bacteria is not in the cleavage of the aromatic ring. Instead, they serve as terminal organisms of the food chain and produce methane from the intermediates formed by the breakdown of benzoic acid by some unknown bacteria.

Further evidence of this conclusion is provided by the fact that the addition of orchlorobenzoate inhibited benzoate degradation, while having no effect on methane production. The unidentified acetate degrader was found to increase its population significantly when acetic acid was added. It is similar in morphology to those seen by other researchers [Pretorius, 1971; Mylorie and Hungate, 1954] but so far has defied isolation in pure

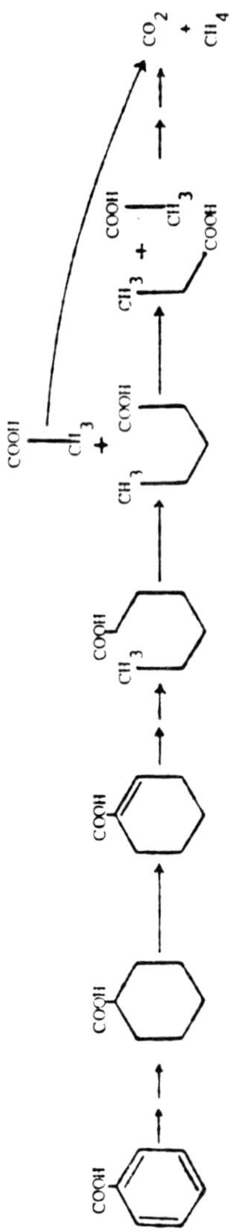

Figure 2.2 Proposed Pathway for Degradation of Benzoic Acid to Methane and Carbon Dioxide (Keith, 1978).

culture. Using data based on the rate of appearance and disappearance of intermediates and the formation of products using labelled C^{14}, Ferry and Wolfe proposed this reaction sequence for the overall conversion of benzoate to methane:

1. Degradation of benzoate to acetate, formate, and H_2:

$$4C_7H_6O_2 + 24H_2O \rightleftarrows 12C_2H_4O_2 + 4CH_2O_2 + 8H_2 \quad (2.2)$$

2. Conversion of acetate to methane and carbon dioxide:

$$12\ C_2H_4O_2 \rightleftarrows 12CH_4 + 12\ CO_2 \quad (2.3)$$

3. Conversion of formate to CO_2 and H_2

$$4\ CH_2O_2 \rightleftarrows 4\ CO_2 + 4H_2 \quad (2.4)$$

4. Reduction of CO_2 with H_2

$$12\ H_2 + 3\ CO_2 \rightleftarrows 3CH_4 + 6H_2O \quad (2.5)$$

5. Sum of Reactions 1-4

$$4C_7H_6O_2 + 18H_2O \rightleftarrows 15\ CH_4 + 13\ CO_2 \quad (2.6)$$

Evans (1977), in a review article on the biochemistry of aromatic compounds in anaerobic environments, proposed the pathway shown in Figure 2.3 for the degradation of benzoate to methane and carbon dioxide. Evans believed that the occurrence of heptanoic acid can be explained by something other than the reduction of cyclohexane carboxylic acid, a reaction which has no biochemical precedent. The proposed intermediate for the precursor of heptanoic acid is 1-methylcyclohexanone, which is converted to heptanoate by the addition of water.

Neufeld (1980) based on the work of Chmielowski (1965a, 1965b, 1966), Keith (1978), and Williams and Evans (1973), proposed a pathway for the degradation of phenols and benzoic acid (see Figure 2.4). This model has as its foundation the detection of cyclohexanecarboxylate and cyclohexene-1-carboxylic acid intermediates by isotopic

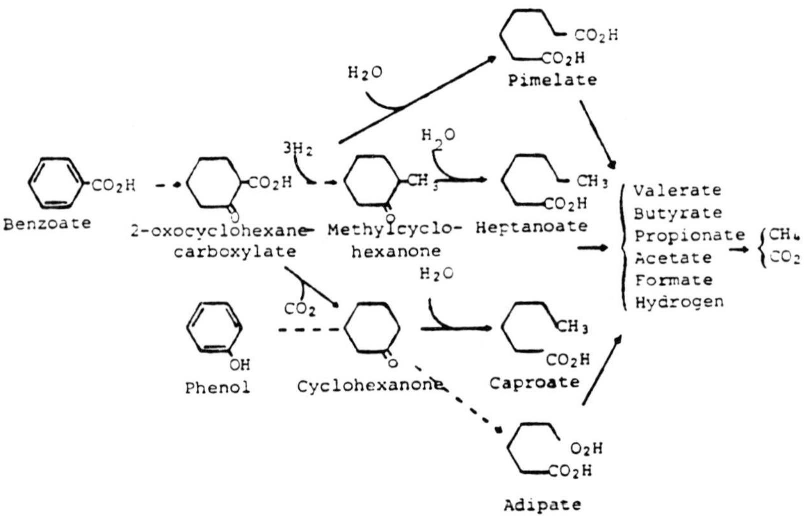

Figure 2.3 Proposed Pathway for the degradation of Benzoate and Phenol into Methane and Carbon Dioxide (Evans, 1977).

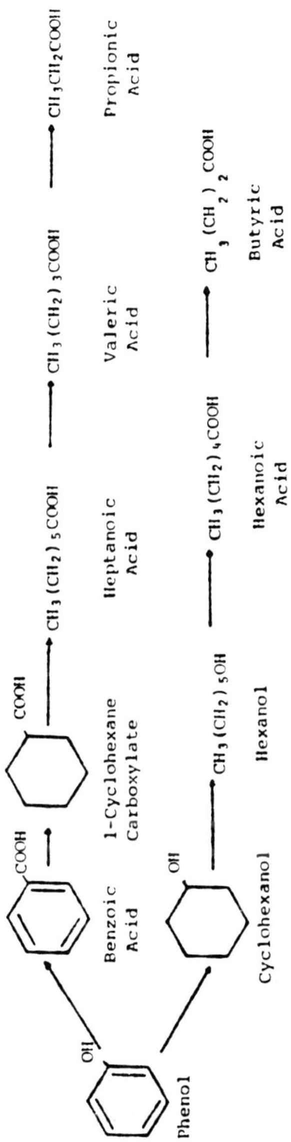

Figure 2.4 Proposed Pathway of Phenol and Benzoic Acid Degradation (Neufeld, 1980).

trapping and thin layer chromatography [Balba and Evans, 1977; Keith, 1973, 1978].

Many other researchers have studied the degradation of aromatic structures by anaerobic processes. Voets, et al. (1976), studied the breakdown of antimicrobial agents, widely used in hospitals and industry. Employing batch reactors operated on either mineral or organic medias, the four antimicrobial phenolic compounds o-phenyl phenol, p-chloro-m-cresol, 5,5′ dichloro 2,2′-dihydroxy-diphenylmethane, and 5-chloro-2-(2,4-dichlorophenoxy) phenol, were tested for susceptibility to bacterial degradation. Of the four compounds, o-phenyl phenol at concentrations of 10 and 40 mg/l was the only one to undergo 100 percent degradation in both mediums. The compounds p-chloro-m-cresol and 5,5′ dichloro-2,2′-dihydroxy diphenyl-methane were not degraded at all, in either media. The compound 5-chloro-2-phenol was degraded by 50 percent in organic medium, but was not reduced to any detectable extent in the mineral media.

Healy and Young (1978), presented evidence that the aromatics catechol and phenol could be degraded by a population of anaerobic bacteria to produce methane and carbon dioxide. Using serum bottle enrichments as batch reactors subjected to incubation in darkness and intermittent shaking, 300 mg/l of either catechol or phenol were fed to a reactor as the sole carbon source. The catechol reactor experienced an acclimation time of 4.5 weeks. Immediately after this, a correlation between the onset of gas production and substrate disappearance was observed. In the case of phenol, 2.5 weeks were necessary for acclimation to take place, with a correlation between the onset of gas production and substrate disappearance observed here also. The reactions can be stoichiometrically described as:

$$\text{Phenol: } C_6H_6O_2 + 4H_2O \rightleftarrows 2.5\ CO_2 + 3.5\ CH_4 \quad (2.7)$$

$$\text{Catechol: } C_6H_6O_2 + 3.5\ H_2O \rightleftarrows 2.75\ CO_2 + 3.25\ CH_4 \quad (2.8)$$

From the above stoichiometry, the amount of gas produced in these experiments was approximately 79 percent and 85 percent of the theoretical gas production for catechol and phenol, respectively.

Healy and Young (1979) showed that eleven aromatics expected to be released in the disposal of a natural refractory compound ligand by heat treatment [Gossett and McCarty, 1976] could be degraded by anaerobic processes. The eleven compounds, tested in small batch reactors at a concentration of 300 mg/l, showed in most cases more than 80 percent conversion of substrate carbon into gas. Acclimation times ranged from a low of two days for syringic acid to a high of 21 days for catechol. Cross acclimation studies were conducted where cultures acclimated to one substance were tested to see if the culture was simultaneously acclimated to another. The results were positive. For example, a culture fed 400 mg/l of vanillic acid could degrade 400 mg/l of catechol with immediate gas production.

Suidan, _et al_. (1979, 1981) utilized an anaerobic granular activated carbon filter to simulate treatment of wastewater bearing catechol, o-cresol, and a mixture of o-cresol and glucose. The test reactors were fed catechol and o-cresol on a batch basis for 3 weeks, after which continuous feeding was initiated at a substrate flow rate of 2 ml/min. The catechol unit run for a total of 617 days. From days 1-333 the reactor was fed a synthetic substrate bearing 200 mg/l oc catechol. During this time, the system functioned as a two step completely mixed carbon adsorption system until a viable bacterial population took hold, after which bioregeneration of the activated carbon was observed. In subsequent steps the catechol bearing feed was increased to final concentrations of 400 mg/l and 1000 mg/l. In both cases the bacterial population continued to degrade the substrate and operate at steady state levels after a relatively short increase in effluent catechol concentrations.

The o-cresol reactor was operated for a period of 732 days. During the first 410 days, at a feed substrate concentration of 256 mg/l, little biological activity was observed. The feed was then supplemented by the addition of 100 mg/l of glucose. It was observed that degradation of glucose was taking place, while the reactor level of o-cresol remained unchanged. Even though o-cresol was not susceptible to degradation, it was found not to be inhibitory to the methanogens breaking down other organic compounds.

Kirsch, et al. (1981) demonstrated the use of an anaerobic semi-continuous stirred tank reactor in the treatment of pentachlorophenol (PCP). In the early stages of testing, PCP exhibited bio-inhibitory characteristics when a soluble concentration of 200 μg/l existed in the reactors. This was reflected by a decrease in gas production. By careful control of the acclimation period, this problem was overcome and the reactors were eventually acclimated satisfactorily to a PCP concentration of 7.6 mg/l, at an SRT of 72.8 days. In a subsequent phase of the experimentation, reactors were established at SRT's of 10, 20, and 40 days, at a PCP feed concentration of 5 mg/l, with normal reactor operation observed.

Food Processing and Agricultural Wastes

Wastes from food processing and agricultural operations have traditionally presented problems in disposal. These wastes are produced in large quantities and are usually of high strength. Numerous accounts from the literature show that anaerobic systems have been widely employed to treat wastes of this type, with encouraging results. The majority of the work done in this area can be divided into four general categories, based on the waste type. These four categories are: sugar refinery wastes, brewery wastes, animal wastes, and food processing operations. A detailed review is beyond the scope of this study and, therefore, each of these general categories will be briefly reviewed. For a more detailed review of the work pertinent to these areas, a list of references grouped in these general categories has been provided at the end of this chapter.

Sugar Processing Wastes

Wastes from sugar refineries have long presented problems in disposal. The effluent wastes are very high in strength, with values of 20,000 mg/l BOD not uncommon. In addition, the wastes have a high carbohydrate content. The ability of anaerobic processes to accept high loading rates and high concentrations of carbohydrates make them ideal for the treatment of these wastes. Various accounts from the literature (see general list of references) have indicated that in most cases, anaerobic systems functioned satisfactory in the breakdown of these wastes and removal of waste BOD were usually in the range of 85-95 percent (see Table 2.3).

Brewery Wastes

Brewery wastes present similar problems to those of sugar processing, namely, high strength. These wastes are composed of the residues of the fermented material after the alcohol has been distilled off. If cereal is used as a raw material, the resulting waste may have a BOD as high as 25,000 mg/l. If molasses is used, the BOD will be higher [Pettet, et al. 1959]. Anaerobic processes have been used to treat these wastes, with good results (see references and Table 2.2). In using an anaerobic system for distillary waste, special precautions must be taken to reduce the inhibitory tendency of sulfates contained in the waste by either dilution or adjusting loading rates.

Animal Wastes

In agricultural operations such as feedlots, the disposal of manure is a tremendous problem that drains manpower and money from other important phases of operation. In addition, recent increases in the cost of fuel have placed additional strain on the industry. The potential for waste disposal, with the added bonuses of energy production, has stimulated much interest in the agricultural field for the use of anaerobic systems. These systems have included everything from crude digesters to elaborate systems involving digestion of the sludge with the digester supernatant being drawn off to be used in the cultivation of algae for protein in roof-top ponds [Lai, et al. 1974]. Others have used the digested material directly as animal feed [Loe and Turk, 1973], and reported that the digested residue had two times the crude protein and ammino acids of the regular wastes and was of comparable quality to that of soybean meal. In addition to the above, anaerobic processes offer the following advantages for treatment of animal wastes.:

1. left over residue makes a good fertilizer with a low pollutant strength if applied to soil.

2. sludge is more amenable to dewatering.

Table 2.2 Operational Data of Studies of the Digestion of Food and Agricultural Wastes. (Fannin, *et al*. 1981 and Chynoweth, *et al*. 1979).

Feed	Temp., °C	Loading, kg VS/m³-day	Retention time, days	Methane yields, m³/kg VS added	Methane Production rate, vol/vol Culture-day	Methane Content mol%	Volatile Solids %	Reference
Fruit & Vegetable Waste	37	--	25	0.24	--	55	90	Lane, (1979)
Piggery waste	30	1.8	20	0.25	0.45	--	--	Van Velsen, (1980)
Piggery waste	30	2.2	--	0.24	0.53	--	--	
Piggery waste	30	3.1	--	0.22	0.68	--	--	
Piggery waste	30	2.7	--	0.23	0.62	--	--	
Piggery waste	30	2.6	15	0.35	0.91	--	--	
Piggery waste	25	2.5	22	0.32	0.80	70	--	Van Velsen, (1979)
Piggery waste	32	3.1	15	0.28	0.87	73	--	
Piggery waste	55	2.7	20	0.26	0.70	59	--	
Swine manure (6-8% TS)	35	1.3	10	0.35	0.46	55-60	60	Ianotti, *et al*. (1980)
Swine Manure (6-9% TS)	35	3.8	15	0.35	1.30	59	60	
Beef Cattle Manure	35	10.0	6	0.16	1.60	--	--	Chen, *et al*. (1980)
Beef Cattle Manure	35	5.5	12	0.23	1.27	--	--	
Beef Cattle Manure	35	3.3	18	0.26	0.86	--	--	
Beef Cattle Manure	55	10.0	6	0.26	2.60	--	--	
Beef Cattle Manure	55	5.5	12	0.28	1.54	--	--	
Beef Cattle Manure	55	3.3	18	0.29	0.96	--	--	
Livestock Manure	55	3.4	20	0.22	0.75	58	44	Hashimoto, *et al*. (1977)
Livestock Manure	55	5.2	12	0.31	1.59	55	53	
Livestock Manure	55	11.4	6	0.23	2.73	53	48	
Livestock Manure	55	15.0	4	0.21	3.14	50	40	
Fresh + dried Manure + Straw with Feed TS								
2.5%	35	2.5	20	0.18	0.44	55	--	Wujcik & Jewell, (1980)
5%	35	5.0	20	0.17	0.82	55	--	
5%	35	10.0	10	0.11	1.10	55	--	
25%	35	17.5	20	0.11	2.93	52	--	
25%	35	35.0	20	0.03	1.05	47	--	
Broiler-Chicken Litter with Feed VS of								
3%	60	6.1	5	0.03	0.20	51	23	Shih & Huang, (1980)
5%	60	10.0	5	0.03	0.28	49	20	
7%	60	22.9	3	0.01	0.32	35	16	
7%	60	14.0	5	0.02	0.28	43	17	
7%	60	10.0	7	0.02	0.23	48	19	
7%	60	7.7	10	0.03	0.23	55	20	
9%	60	18.4	5	0.02	0.35	42	14	
Beef Cattle Waste	55	3.4	20	0.22(VS)	0.75	58	44.2	Hashimoto, *et al*. (1978)
		5.2	12	0.31	1.59	55	52.8	
		11.4	6	0.23	2.73	53	46.1	
		15	4	0.21	3.14	50	39.8	

Table 2.1 (Continued)

Feed	Temp., °C	Loading, kg VS/m^3-day	Retention time, days	Methane yields, m^3/kg VS added	Methane Production rate, vol/vol Culture-day	Methane Content mol%	Volatile Solids %	Reference
Dairy cattle waste	35	7.0	12	0.049	0.34	65	21	Smith, (1978)
Dairy Cow manure	32.5	27.8	2.5	0.007	0.182	59	8.6	Morris, et al. (1978)
	32.5	12.1	5	0.032	0.385	65	23.2	
	32.5	17.4	5	0.063	1.10	61	18.8	
	32.5	3.49	10	0.177	0.616	66	35.0	
	32.5	8.70	10	0.142	1.233	64	29.0	
	32.5	1.74	20	0.233	0.405	66	37.8	
	32.5	4.35	20	0.178	0.773	63	35.1	
	32.5	1.15	30	0.237	0.273	65	39.8	
	32.5	2.91	30	0.198	0.575	63	37.6	
Bean blanching wastes	35	6.6	30	----	----	--	80%(COD)	Van denBerg & Lentz, (1978)
Pear peeling wastes	35	3.0	10	----	----	--	90%(COD)	
Potato peeling wastes	35	1.6	15	----	----	--	80%(COD)	
Rum stillage	35	2.4	18	----	----	--	80%(COD)	
Sugar beet plant wastewater	30	3.5(TOC)	0.12	0.61*(TOC)	2.14*	--	75(TOC)	Heertjes and Van der Meer, (1978)
	30	4.5(TOC)	0.13	0.80*(TOC)	3.60*	--	85(TOC)	
	30	5.5(TOC)	0.18	0.79*(TOC)	4.35*	85	95(TOC)	
Whey	22-25	1.9(COD)	----	0.45-0.50(COD)	0.86-0.95	--	97-98% (COD)	Hakansson, et al. (1977)

* Total gas yields and production rates

TABLE 2.3 Results of Screening Studies of Selected Bacteria for Their Capability to Degrade γ-BCH Under Anaerobic Conditions [Jagnow, *et al*. 1977].

Bacterium	Degradation	Bacterium	Degradation
Clostridium butyricum	+++	*Citrobacter freundii*	+++
C. pasteuriamun	+++	*Escherichia coli*	++
		Enterobacter aerogenes	++
Bacillus polymyxa	++	*E. cloacae*	+
B. macerans	++	*Serratia marcescens*	+
B. laterosporus	+	*Proteus mirabilis*	+
Bacillus alvei	±	*Proteus vulgaris*	±
B. circulans	±		
B. brevis	-	*B. lentus*	-
B. cereus	-	*B. licheniformis*	-
B. coagulans	-		
Lactobacillus casei	-	*Leuconostoc dextranicus*	-
L. plantarum	-		
L. brevis	-	*L. mesenteroides*	-
Paracoccus dentrificans	-	*Propionibacterium Shermanii*	-

+++, ++, + = strong to moderate degradation

± = ambiguous degradation

- = no degradation

3. sludge produced is inert and nuisance free.

Some problems do exist with the use of anaerobic systems in treatment of animal wastes. In addition to those of scum formation and clogging of gas ports, NH_4 toxicity and reduced methane production are major problems. Ammonia toxicity is a major problem when dealing with fresh waste that contain a large amount of urine. Lapp, et al. (1975) and Kroeker, et al. (1975) have suggested that this problem can be offset by proper acclimation of the methane formers, and cation antagonism. Most of the data available (see references at end of chapter and Table 2.2) has shown the gas production is not produced in the amounts expected. Additional work is needed to explain this phenomenon.

Food Processing

Anaerobic processes have been applied to a wide array of food processing operations. Some of the types of operations covered include alewife and tuna processing, pear and pineapple wastes, citrus wastes, and bean blanching wastes. Because of the diversity of the information available in this area, a discussion has not been included. For more information one should consult the list of references at the end of this chapter under Food Processing. Some experimental data from past research is presented in Table 2.2.

Soaps and Detergents

Much of the public's concern over biodegradability stems from the subject of soaps and detergents and their resulting effects on receiving waters. While much work has been done on aerobic breakdown of soaps and detergents, little had been done on anaerobic processes. Vath (1964) conducted some preliminary work with several linear anionic and nonionic ethyxylated surfactants. Using batch reactors operated between 22-28°C with initial surfactant concentrations of 20 and 100 mg/l, Vath found that ethoxamers derived from both linear secondary alcohols, and α olefin alkylphenols, underwent degradation as observed by a loss of surfactant properties. The degradation proceeded by a similar mechanism to that of aerobic degradation, differing only in the hydrogen acceptor employed (see Figure 2.5).

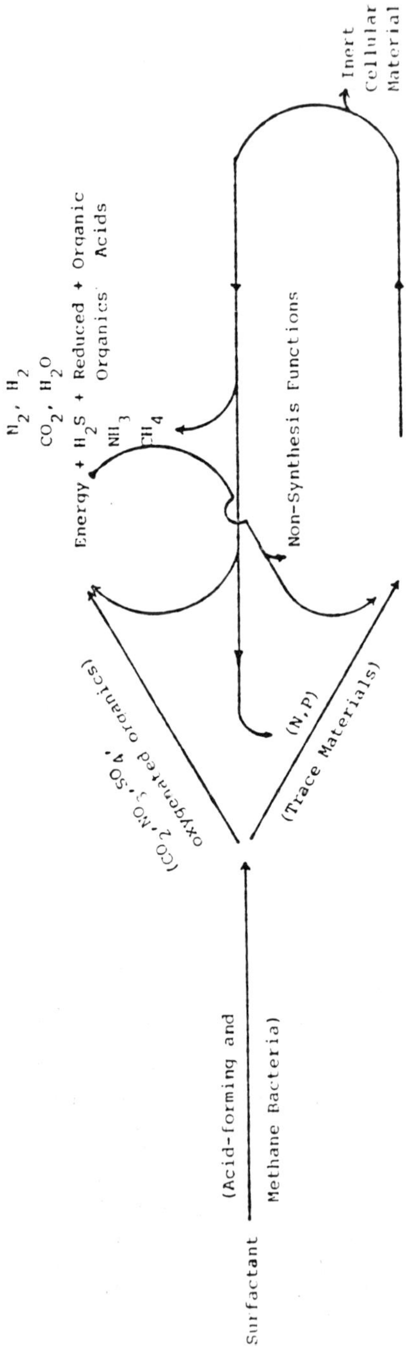

Figure 2.5 Pathway for the Anaerobic Biodegradation of Surfactants (Vath, 1960).

Saldick (1974) studied the breakdown of cyanuric acid (S-triazine 2,4,6-triol), which is the end product of the partial breakdown of sodium or potassium dichloroisocyanurate and trichloroisocyanuric acid, both of which are used in the manufacture of household cleaners and dishwater detergents. Using primary settled domestic sewage under anaerobic conditions, to which 10 mg/l of cyanuric acid had been added, Saldick found a 25-50 percent reduction of the cyanuric acid concentration in 48 hours and complete disappearance in 72-96 hours. Extending this experimentation, employing various substrates and initial O_2 concentrations, Saldick concluded that cyanuric acid was broken down more readily by anaerobic means than by aerobic processes. In anaerobic media the overall degradation occurs by a biogically catalyzed hydrolysis of the type:

$$C_3H_3N_3O_3 + 3H_2O \rightleftarrows 3CO_2 + 3NH_3 \qquad (2.9)$$

Barth and Bunch (1979) conducted research on the breakdown of sodium nitrilotriacetic acid (NTA), a possible substitute for phosphates in detergents. They resolved to settle conflicting reports on whether NTA was amenable to anaerobic breakdown. Working with a primary sludge feed system at a concentration of 20 mg/l NTA, it was concluded that after a period of 120 days no appreciable degradation had occurred. However, when the procedure was repeated using a 50/50 mixture of primary sludge and activated sludge acclimated to NTA, concentrations in the digester decreased to near zero. To verify that the digester was operating properly, a slug dose of 20 mg/l was added to the digester (along with continuation of the 20 mg/l sludge feed) and a near complete removal of NTA from the digester was recorded 15 days later.

Trisodium carboxymethloxysuccinate (CMOS) is a new detergent builder that contains neither phosphorous nor nitrogen, as the most commonly used builders do. Klein (1974) showed that CMOS could be broken down anaerobically using 7 cm diameter soil columns loaded with 35 cm of sandy loam and purged with CO_2 to remove O_2. Using septic tank effluent as feed, 10, 40, and 100 mg/l concentrations of CMOS were completely removed after a 2 week acclimation period. Vicarro (1977) expanded on this subject by using C^{14} labelled CMOS to study the extent of the biodegradation reaction. In a one liter laboratory anaerobic digester, run at a detention time of 32 days and an organic loading rate of .00190 lb. volatile solids/day, it was found that a concentration of 20 mg/l CMOS would undergo degradation with CO_2 and CH_4 as virtually the only end products. After the acclimation time, the removals were rapid and nearly complete, with 97 percent removals being achieved.

Pharmaceutical Wastes

Wastes from pharmaceutical manufacturing operations are ideal for treatment by anaerobic processes, due to the high COD and low suspended solids characteristics of these wastes. Jennett and Dennis (1975) employed upflow anaerobic filters in the treatment of a pharmaceutical waste. The influent waste had a high COD value, attributed to the presence of methanol in the wastes. Removal efficiencies were reported to be between 93 and 98 percent, at a loading rate of 13.8 to 220 lb COD/day/ 1,000 cubic feet. Although it was admitted that methanol is a readily degradable substance, some important results were attained. Attempts to disrupt filter performance by the addition of sharp increases in the organic loading rate resulted in failure. Probably the most significant result of this experiment was the demonstration that solids were easily kept in the system at a manageable level, with little loss to the effluent.

Sachs, et al. (1978) used an anaerobic filter in the treatment of wastes from a synthesized organic chemical manufacturing plant. The filters were started on a methanol feed solution containing 2,000 mg/l COD. After steady state conditions were reached, the feed was switched to the plant waste diluted down to 2,000 mg/l COD, at an organic loading rate of 34.9 pound COD/1,000 cubic feet/ day. At steady state conditions the COD removals ranged from 70 to 80 percent. A three-fold increase in influent COD concentration resulted in disruption of filter performance, evident by a 18 percent reduction in COD removal. When operated at the 2,000 mg/l COD level, the filters provided 33 percent better removal of pound COD/MG/day and gave a less colored effluent than the existing treatment system.

Pesticides

The fact that pesticide residues persist in the environment is well-established. Numerous cases exist where residues have been detected more than a decade after application. These residues are toxic to many organisms and have a potential for bioaccumulation. Many scientists believe that if a compound is resistant to aerobic conditions, it will not degrade at all under anaerobic conditions. The work to date on anaerobic breakdown of pesticides would lead us to think otherwise. It is well documented that anaerobic conditions existing in flooded rice fields facilitate degradation of pesticides better than aerobic conditions. Studies dealing with specific pesticides are described below.

The organochlorine insectide 1,2,3,4,5,6-hexachlorocyclohexane (γ-HCH), better known as lindane, jacutin, gammexane or γ benzene hexachloride (γ-BHC), and its related isomers, α-BHC, β-BHC, and δ-BHC, are all on the EPA priority pollutants list. Of the above, only the γ-BHC isomer is effective as an insecticide. The other isomers, α-BHC, β-BHC, and δ-BHC, are formed in the manufacturing of γ-BHC, and have no value in pest control. They are very resistant to biological breakdown and yet they have been applied to agricultural areas in quantity in unrefined lindane. Many countries, including Japan, have banned the use of lindane altogether.

Hill and McCarty (1967) in their study on anaerobic degradation of chlorinated pesticides, found that under anaerobic conditions at 35°C, lindane was degraded much more rapidly than under aerobic conditions. However, at 20°C this difference was less pronounded. Prior to this, it was thought that none of the chlorinated hydrocarbons could be used as a carbon source for microbial metabolism. It was observed in the course of these experiments that degradation best occurred under conditions of high biological activity. Reaction rates appeared to be a function of the concentration of sludge and lindane. At high concentrations of both sludge and lindane the reaction rate followed first order kinetics, but when diluted solutions of sludge were used at low concentrations of lindane a reaction rate of zero order was observed.

MacRae, et al. (1967) and his colleagues studied the degradation of all four isomers of benzene hexachloride in submerged soils columns. Since any available O_2 present in the soil was soon utilized, the submerged soil quickly became anaerobic. Applications of the four isomers to the tubes were carried out at three times the expected dosage of commercial applications. It was found that no evidence for prolonged existence of any isomer existed, when the columns were analyzed for benzene hexachloride. All four isomers were degraded at approximately the same rates, and degradation was faster in unsterilized soil than in sterilized soil, again suggesting the involvement of biological activity. To confirm these results, C^{14} labelled γ-BHC was introduced into the soil and the soil surface measured for release of labelled CO_2. The results obtained demonstrated that biological activity was indeed taking place, resulting in release of labelled CO_2.

MacRae, et al. (1967), in further experiments isolated a bacterium identified as *Clostridium sp.*, that was capable of degrading lindane in phosphate buffer using the lindane as the sole carbon source. It was shown that an initial concentration of 3.67 mg/l of lindane could be degraded to 0.02 mg/l in 27 hours. Ability of the organism to degrade lindane was also provided by the release of the covalently linked chlorine of lindane as free chlorine ion. For the amount of lindane degraded, 75 percent of the theoretical amount of chloride expected to be released was measured in the culture.

Sethunathan, et al. (1969) also worked with a species of *Clostridium*, in attempts to identify degradation intermediates. The compound γ-pentachlorocyclohexene, a probable intermediate of the direct dehydrochlorination of γ-BHC, was not detected. This indicated that another mechanism was operating. Based on the similarity of the factors affecting lindane and DDT degradation and the fact that the *Clostridium sp.* could convert DDT to DDD by a reductive chlorination, it was suggested that γ-BHC is degraded by a similar pathway. However, scientific data to support this hypothesis is unavailable.

Newland, et al. (1969) obtained evidence indicating that degradation of lindane occurs faster under anaerobic conditions than under aerobic conditions. In simulated lake impoundments, anaerobic degradation of lindane was more complete (approximately 90 percent vs 15 percent) and began earlier (264 hours vs 840 hours) than aerobic degradation. In attempts to define degradation products, gas-liquid and thin layer chromatography were performed on samples from the impoundments. Both methods detected the presence of the α and δ isomers of BHC as the products of a biological meditated isomeric conversion. While this result was unexpected, it is acceptable on the grounds of thermodynamics since the γ isomer is the least thermodynamically stable of the four isomers (γ-BHC < α-BHC δ-BHC < β-BHC). Since, in this reaction, no carbon of BHC is utilized, the reaction appears to be of no metabolic value. Instead this conversion may serve as a response by the organism(s) to reduce the toxicity in its environment.

Jagnow, *et al*. (1977) performed screening studies on selected species of bacteria and recorded their capability to degrade γ-BHC anaerobically. The results are presented in Table 2.3. It is evident that the *Clostridia* were highly active, with some members of the *Bacillacae* and *Enterobacteriacea* also performing quite well. Nearly complete dechlorination in 4-7 days at a concentration of 2 mg γ-BHC per 200 ml. of medium was observed for the three species, *Clostridium butyricum*, *Clostridium pasteurianum*, and *Citrobacter freundii*. The use of gas chromatography aided in the identification of one reaction intermediate, γ-tetrachloro-cyclohexene. In addition tri and tetrachlorobenzenes were detected but not in sufficient enough quantities to appear as major metabolites in the destruction of γ-BHC.

DDT

DDT (dichlorodiphenyl-trichloroethane) has the chemical formula 1,1,1-trichloro-2,2-bis-(4-chlorophenyl)ethane. It was invented by the Swiss during World War II, and marketed under the trade names Gerasol, Guesarol, and Necocid. Because of its highly intrinsic insecticidal value it has found widespread use in the control of a variety of pests. DDT is known to persist for many years in soil. Past evidence has decisively shown that DDT is degraded faster under anaerobic conditions than aerobic conditions, and the degree of degradation is inversely proportional to the amount of oxygen in the system [Ko and Lockwood, 1968; Hill and McCarty, 1967; Castro and Yoshida, 1971]. The extent of degradation is slight, with most of the DDT being reductively dechlorinated to DDD, which is somewhat less toxic, but more bioresistant than DDT (see Figure 2.6 for pathway). Castro and Yoshida (1971) have found that DDD will degrade in flooded fields but at a much slower rate.

Guenti and Beard (1968) first provided unequivocal evidence that DDT is degraded faster under anaerobic conditions than aerobic conditions. In an anaerobic soil system, the conversion of DDT to DDD was achieved, with ultimate minute traces of six other degradation products. Addition of organic material in the form of alfalfa increased the conversion rate. With the alfalfa present, less than 1 percent of the original amount of DDT applied to the soil was recovered after 12 weeks of incubation. The aerobic soil system, on the other hand, showed 75 percent of the added DDT after a 6 month incubation period. Later research by Cuenzi, *et al*. (1971) showed that other organic substrate such as rice, straw and cellulose can also enhance the conversion of DDT to DDD in anaerobic soils. Organic material added to non-flooded fields under aerobic conditions has no effect on this conversion.

The ability to degrade DDT to DDD has been demonstrated by many organisms. Previous research has shown that *Escherichia coli*, *Aerobacter aerogenes* [both by Mendel and Walter, 1966], *Serratia marcescens* [Stenersen, 1965] and *Proteus vulgaris* [Barker, *et al*. 1965] all are capable of carrying out this conversion. Enzymes involved in this reaction have not been identified as yet, but the site of this reaction in *E. coli* was found to be in the membranous fraction of cells [French and Hoopingarner, 1970].

Other Organochlorine Pesticides

Besides lindane and DDT, most other organochlorine pesticides are thought to be fairly stable. In some instances, however, they have been shown to undergo degradation by anaerobic processes. Hill and McCarty (1967) tested the compounds aldrin, dieldrin, heptachlor, endrin, and heptachlor epoxide in biologically active anaerobic sludge. It

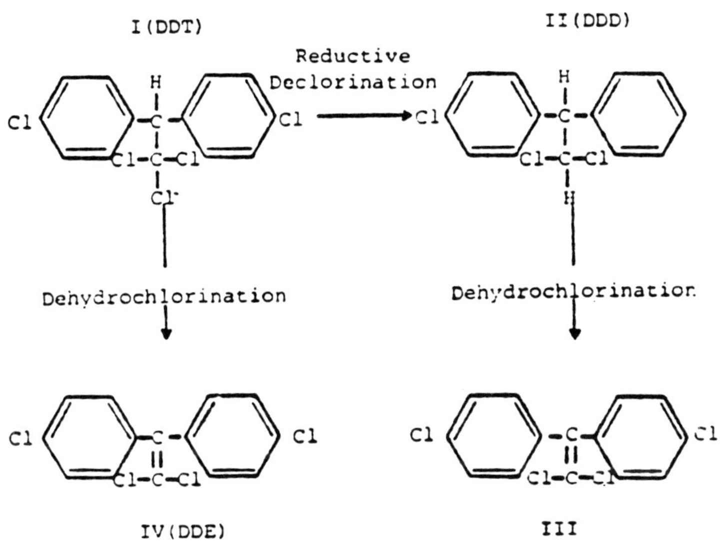

Figure 2.6 The Structure of DDT and some Related Substances (Kallman & Andrews, 1963)

was found that heptachlor and aldrin were degraded at a temperature of 35°C to unidentified products. Castro and Yoshida (1971) showed that methoxychlor and heptachlor degraded faster in anaerobic conditions than aerobic conditions (see Figure 2.7). It was shown that this degradation was the result of biological activity and not volatilization, as other workers had thought. The pesticides aldrin, endrin, dieldrin, and chlordane were shown to be relatively resistant under both anaerobic and aerobic conditions.

Other Priority Pollutants

Bouwer, _et al_. (1981) studied the anaerobic breakdown of two concentrations of the priority pollutants chloroform, trichloroethylene, tetrachloroethylene, and bromodichloromethane. Another compound, dibromochloromethane, was also studied. It was found that chloroform degradation occurred completely at starting concentrations of 16 and 34 µg/l. An initial concentration of 157 µg/l resulted in a less pronounced degradation, and occurred very slowly. Degradation of chloroform was only observed in cultures seeded with methanogenic bacteria. The authors proposed a possible degradation route consisting of both biological and chemical mechanisms. Chloroform combines with a sulhydryl group from the sodium sulfide (added to the media to maintain reducing conditions), forming a thiol. This resulting thiol is then attacked by microorganisms and degraded further. Since no attempt was made to detect intermediates or end products, this pathway is hypothetical.

Both trichloroethylene and tetrachloroethylene showed no conclusive evidence that degradation took place. This trend was observed at all starting concentrations. While it appeared that a small amount of decomposition was occurring, the results were not conclusive.

Degradation of bromodichloromethane at initial concentrations of 16, 35, and 161 µg/l was observed to take place quickly (2 weeks). Similarly, dibromochloromethane underwent rapid degradation at initial concentrations of 19, 45, and 205 µg/l. The removal of both bromodichloromethane and dibrochloromethane occurred in both the control and seeded cultures, which suggests a chemical degradation route. Biological activity may also play an

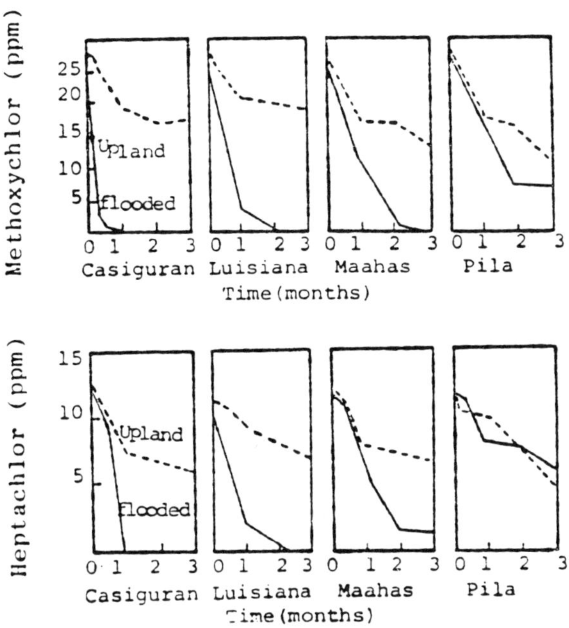

Figure 2.7 Degradation of Methoxychlor and Heptachlor in four soil types (Castro and Yoshida, 1971).

important role in this degradation, since the degradation rates were faster in the seeded cultures than in the control. A mechanism similar to that of chloroform degradation was suggested, where substitution of a sulfhydral group for the bromine atom(s) produces a thiol which is then degraded by some biological mechanism which may increase the reaction rate.

Other Selected Organic Compounds

It has been established that some of the higher molecular weight hydrocarbons that make up oil are decomposible by anaerobic bacteria. Novelli and Zobell (1944) reported that a species of *Desulfovibrio* could degrade eicosane, docosane, paraffin oil, and paraffin wax in the absence of other types of organic matter. Of the hydrocarbons used, part was found to be utilized in cell synthesis, part was oxidized into unknown compounds, while the remainder was broken down into lower weight hydrocarbons.

Rosenfeld (1947) provided further evidence that members of the genus *Desulfovibrio* catalyze the anaerobic oxidation of hydrocarbons. As in the previous account, it was found that the longer chain aliphatic constituents were the most readily attacked. Insoluble fatty acids were detected as reaction intermediates but were transitory in nature and underwent further degradation. It was suggested that a dehydrogenase system is the mechanism by which *Desulfovibrio* is able to utilize hydrocarbons.

Shelton and Hunter (1975) demonstrated that anaerobic decomposition of oil occurred naturally in bottom deposits. It had been previously thought that this reaction proceeded slowly, if it took place at all. This hypothesis was based on the assumption that reservoirs of oil underground would be nonexistent, having been long transformed by bacteria. It is now thought that while bacteria may have initially attacked the oil deposits, the accumulation of toxic metabolics severely limited their growth. While it was acknowledged that some degradation of hydrocarbons was taking place, in their study most of the oil lost from the sediments was attributed to diffusion of the hydrocarbons into water above the sediments rather than by direct biological breakdown. It was postulated that some biochemical action of the microorganisms in the soil converted the hydrocarbons into more soluble intermediates which escaped into the overlying water.

Compounds Nonamenable to Biodegradation

Most organic compounds are biodegradable to an extent. While they may not be degraded to the extent one would like, nevertheless, there exist bacteria that are capable of carrying out slight changes in the nature of these molecules. There do exist however, classes of extremely refactory organic compounds that because of their toxic, complex, or inert characteristics are resistant to microbial attack. This condition may manifest itself in the individual molecules or occur only above certain threshold concentrations. The resistance encountered can be the result of many factors, which include, solubility, molecular size, amount of tertiary branching, and the number, nature, and position of the substituents. There exist general rules to determine the relative biodegradability/nonbiodegradability of a compound. A few of them are as follows: [Painter, 1974; SCS Engineers, 1974]:

1. Materials that can pass through cell membranes are more readily available to the microbes and are degraded faster.

2. Microbes prefer non-aromatic or cyclic aromatics over aromatics. The presence of a substituent in the benzene ring usually increases biodegradability.

3. Soluble compounds are more easily degraded than insoluble.

4. A high degree of branching imparts greater resistance to degradation.

5. Dispersed compounds provide more surface area for attack, and therefore, degrade better.

6. Compounds with unsaturated bonds are degraded more readily than saturated compounds.

7. Such materials are alcohols, aldehydes, acids, esters, amides, and amino acids are preferred over the corresponding alkanes, alkenes, ketones, dicarboxylic acids, nitriles, amines, and chloroalkanes.

Future Areas of Research and Development

From the literature review presented above it is evident that many large gaps exist in the application of anaerobic processing to the treatment of industrial wastes. Little or no work has been published involving the treatment of such groups as chlorinated alkanes, phthalate esters, or polychlorinated biphenyls by anaerobic processes. In addition the basic pathways for the production of methane from many organic substances is not yet well understood and needs to be developed further.

Much of the future work done on anaerobic systems will revolve around process modifications. By manipulation of such factors as SRT, reactor types, and loading rates, process efficiencies will be increased towards the optimal performance level. Other chemical factors, especially those of temperature and pH, need to be carefully studied to maximize the growth rate of microorganisms involved in the biological degradation of organic compounds. Intensive experimentation in the above-mentioned areas will increase the level of technology in the treatment of industrial wastes by anaerobic processes to a much higher plateau than is realized today.

A promising area of future research that relies on the creation of new technology instead of modification of old technology is that of genetic engineering. Applications of this new innovative technology are opening new frontiers in many areas of science, including the wastewater treatment field. While it is highly probable that given a sufficient amount of time, bacteria will acquire the ability to degrade most of the new industrial pollutants on prolonged exposure, it may be possible to significantly speed up the evolutionary process. By selective use of naturally occurring breeding processes, the production of new characteristics in bacteria can be completed many times faster than if the change occurred by natural evolution. Using the techniques of true genetic engineering, the production of new characteristics in bacteria can be increased by an order of magnitude 10^{8-9} times faster than natural evolution [Widdus and Ault, 1974].

The genetic manipulation of organisms can be brought on by many techniques. Genetic material can be transferred from one organism to another via viruses, plasmids, bacteriophages, or even protoplasts. Geneticists are only beginning to realize the potential of genetic engineering using these methods. Some of the possibilities include membrane changes, adding new proteins to cells, and making additional enzyme systems operational. Conversely, properties can be deleted from the genetic makeup by careful introduction of mutagens.

As stated above, plasmid transfer is one possible method of changing the genetic makeup of a particular organism. Plasmids are small bits of double stranded DNA that contain specific information for the production of enzymes needed by the cell in cases of extreme stress. An average plasmid can contain approximately 20-30 genes. By a highly specialized transfer process, the strands of DNA from a donor plasmid can be incorporated into another organism's genetic inventory. The exchanged DNA is duplicated to form a double strand. Conjugation between two organisms can result in the exchanging of this new material, resulting in the eventual establishment of the new genes in the population [Chakabarty, 1976].

This technique was the one employed by Chakabarty and his colleagues at General Electric in the genetic engineering of a microbe that could degrade oil [Chakabarty, 1976]. This was a particularly difficult feat since oil is composed of a variety of hydrocarbons and most microbes can degrade only one specific hydrocarbon. To produce an organism that could degrade oil, plasmids from a variety of organisms, each containing the genetic information for the production of enzymes necessary to degrade a different hydrocarbon, had to all be incorporated into one cell. Growth experiments were performed to test the success of the transplants and it was found that the "new" organism grew much faster than single plasmid ones, indicating that it was using all the individual hydrocarbons contained in oil, simultaneously.

Another possible use of this technique in wastewater treatment involves increasing organisms resistance to toxicants. A case of this type has already been demonstrated by Chakabarty and his colleagues [Chakabarty, 1976]. By transforming the MER plasmid from cells of *Pseudomonas oleovorans* to *Pseudomonas putida*, the resistance of *P. putida* to mercury could be increased by a factor of 25 to 35 times. It is thought that the MER plasmid increases the cell's ability to bind mercury with intracellular protein which is retained in the cell.

While genetic engineering is one solution to solving the problems of treating industrial pollutants it must be emphasized that it is not the only way. As stated before, process modifications while not as "trendy" as the area of genetic engineering, will play a large part in increasing the acceptability of anaerobic processes.

REFERENCES

Balba, M.T. and Evans, W.C., "The Methanogenic Fermentation of Aromatic Substrates." *Biochem. Soc. Transactions*, Vol. 5 (1977).

Barker, P.S., Morrison, F.O., and Whitaker, R.S., "Conversion of DDT to DDD by *Proteus vulagris*, a Bacterium Isolated From the Intestinal Flora of a Mouse." *Nature* 205, 621 (1965).

Barth, E.F. and Bunch, R.L., "Biodegradation and Treatability of Specific Pollutants." *EPA*-600/79-034 (1979).

Bouwer E.J., Rittman, B.E., and McCarty, P.L., "Anaerobic Degradation of Halogenated 1- and 2- Carbon Organic Compounds." *Environmental Science and Technology*, 15, 5, 596 (1981).

Bunch, R.L. and Chambers, C.W., "A Biodegradability Test for Organic Compounds." *J. Wat. Poll. Con. Fed*. 39, 2, (1967).

Castro, T.F. and Yoshida, T., "Degradation of Organochlorine Insecticides in Flooded Soils in the Philippines." J. Agr. Food Chem., 19, 1168 (1971).

Chakabarty, A.M., "Which Way Genetic Engineering." Industrial Research, Jan., 45 (1976).

Chmielowski, J. and Kusznik, W., "Methane Fermentation of Some Phenolic Wastewaters." Zesz. Nauk. Politech. Slaska. Inz. (Pol.), 8, 95 (1965).

Chmielowski, J. and Wasilewski, W., "A Study of the Dynamics of Anaerobic Decomposition of Some Phenols in Methane Fermentation." Zesz. Nauk. Politech. Slaska. Inz. (Pol.), 8, 95 (1965).

Chmielowski, J., "Biochemical Degradation of Some Phenols During Methane Fermentation." Zesz. Nauk. Politech. Slaska. Inz. (Pol.) 8, 97 (1965).

Chynoweth, D.P., Ghosh, S., and Packer, M.L., "Anaerobic Processes. J. Wat. Poll. Con. Fed., 51, 6, 1200-1214 (1979).

Clark, F.M. and Fina, L.R., "The Anaerobic Decomposition of Benzoic Acid During Methane Fermentation." Arch. Biochem., 36 26-32 (1952).

Dutton, P.L. and Evans, W.C., "The Metabolism of Aromatic Compounds by Rhodopseudomonas palustris." Biochem. Jour., 113, 525 (1969).

Dutton, P.L. and Evans, W.C., "Dissimilation of Aromatic Substrates by Rhodopseudomonas palustris." Proc. of the Biochem. Soc., 104, 30P (1967).

Evans, W.C., "Biochemistry of the Bacterial Catabolism of Aromatic Compounds in Anaerobic Environments." Nature, 270, Nov. 3 (1977).

Fannin, K.F., Conrad, J.R., Jerger, D.E., Srivastava, V., Ghosh, S., and Chynoweth, D.P. "Anaerobic Processes." J. Wat. Poll. Con. Fed., 53, 6, 711-720 (1981).

Ferry, J.G. and Wolfe, R.S., "Anaerobic Degradation of Benzoate to Methane by a Microbial Consortium." Arch. Microbiol., 107, 33-40 (1976).

Fina, L.R. and Fiskin, A.M., "The Anaerobic Decomposition of Benzoic Acid During Methane Fermentation. II. Fate of Carbons One and Seven." Arch. Biochem., 91, 163-165 (1960).

French, A.L. and Hoopingarner, R.A., "Dechlorination of DDT by Membranes Isolated from Escherichia coli." J. Econ. Entomol., 63, 756 (1970).

Gilbert, P.A. and Lee, C.M. "Biodegradation Test: Use and Value." Taken from the Biotransformation and Fate of Chemicals in the Aquatic Environment, Proceedings of Workshop held at the Univ. of Mich. Bio. Station, Aug. 1979. (Amer. Society for Microbiology).

Glover, S.W., "Aspects of Genetic Engineering in Microorganisms." Advances in Microbial Physiology, Vol. 18, 1978 (Academic Press).

Gossett, J.M. and McCarty, P.L., "Heat Treatment of Refuse for Increasing Anaerobic Biodegradability." In Biochemical Engineering-Energy. Renewable Resources and New Foods. Vol. 72, No. 158, pp. 64-71 (American Institute of Chemical Engineers Symposium Series, 1976).

Guenzi, W.D., Beard, W.E., and Viets, F.G., "Influence of Soil Treatment on Persistence of Six Chlorinated Hydrocarbon Insecticides in the Field." Soil Sci. Soc. Amer. Proc. 35, 910 (1971).

Guenzi, W.D., Beard, W.E., "Anaerobic Conversion of DDT to DDD and Aerobic Stability of DDT in Soil." Soil Sci. Soc. Amer. Proc., 32, 522 (1968).

Guyer, M. and Hegeman, G.D., "Evidence for a Reductive Pathway for the Anaerobic Metabolism of Benzoate." J. Bact., 99, 906-907 (1969).

Healy, J.B. and Young, L.W., "Anaerobic Biodegradation of Eleven Aromatic Compounds to Methane." Appl. and Environ. Microbiol., 38, 1, 84-89 (1979).

Healy, J.B., Jr. and Young, L.Y., "Catechol and Phenol Degradation by a Methanogenic Population of Bacteria." Appl. Environ. Microbiol., 35, 216-218 (1978).

Hill, W.D. and McCarty, P.L. "Anaerobic Degradation of Selected Chlorinated Hydrocarbon Pesticides." J. Wat. Poll. Con. Fed., 39, 8, 1259-1277 (1967).

Hopwood,, A.P. and Downing, A.L., "Factors Affecting the Rate of Production and Properties of Activated Sludge in Plants Treating Domestic Sewage." J. Proc. Inst. Sew. Purif., 435-448 (1965).

Hutton, D.G., "Removal of Priority Pollutants." Industrial Wastes, March/April, 1980.

Jagnow, G., Haider, K., and Ellwardt, P., "Anaerobic Dechlorination and Degradation of Hexachlorocyclohexane Isomers by Anaerobic and Faculative Anaerobic Bacteria." Arch. Microbiol., 115, 285-292 (1977).

Jennett, J.C. and Dennis, N.D., "Anaerobic Filter Treatment of Pharmaceutical Waste." J. Wat. Poll. Con. Fed., 47, 1, 104 (1975).

Kallman, B.J. and Andrews, A.K., "Reductive Dechlorination of DDT to DDD by Yeast." Science, 141, 1050-1051 (1963).

Keith, C.L., Bridges, R.L., Fina, L.R., Iverson, K.L., and Cloran, J.A., "The Anaerobic Decomposition of Benzoic Acid During Methane Fermentation IV. Dearomatization of the Ring and Volatile Fatty Acids Formed on Ring Rupture." Arch. Microbiol., 118, 173-176 (1978).

Keith, C.L., Dissertations Abstr. Intern., B33, 3214-3215 (1972).

Kirsch, E.J., Grady, C.P.L., and Waukasch, R.F., "Protocol Development for the Prediction of the Fate of Organic Priority Pollutants in Biological Wastewater Treatment Systems." U.S. Environmental Protection Agency, March 1981.

Klein, S.A. and Jenkins, D., J. Wat. Poll. Con. Fed., 46, 2107 (1974).

Ko, W.H. and Lockwood, J.L., "Conversion of DDT to DDD in Soil and the Effect of these Compounds on Soil Microorganisms." Can. J. Microbiol., 14, 1069 (1968).

Kugelman, I.J. and McCarty, P.L., "Cation Toxicity and Simulation in Anaerobic Waste Treatment." J. Wat. Poll. Con. Fed., 37, 97 (1965).

MacRae, I.C., Raghu, K., and Bautista, E.M., "Anaerobic Degradation of the Insecticide Lindane by Clostridium sp." Nature, 221, 859-860 (1969).

MacRae, I.C. Raghu, K., and Castro, T.F., "Persistence and Biodegradation of Four Common Isomers of Benzene Hexachloride in Submerged Soils." J. Agr. Food Chem., 15, 5, 911-914 (1967).

McCarty, P.L., Anaerobic Waste Treatment Fundamentals, Part I. Public Works. N.Y., Sept. 1964, p. 107.

McCarty, P.L., Jeris, J.S., and Murdoch, W., Individual Volatile Acids in Anaerobic Treatment." J. Wat. Poll. Con. Fed., 35, 12, 1501-1516 (1963).

Mendel, J.L. and Walton, M.S., "Conversion of p,p-DDT to p.p-DDD by Intestinal Flora of the Rat." Science, 151, 1527 (1966).

Mylroie, R.L. and Hungate, R.E., "Experiments on the Methane Bacteria in Sludge." Canad. J. Microbiol., 1, 55-64 (1954).

Neufeld, R.D. and Mack, J.D., "Anaerobic Phenol Biokinetics." J. Wat. Poll. Con. Fed., 52, 9 (1980).

Newland, L.W., Chesters, G. and Gerhard, B.L., "Degradation of -BHC in Simulated Lake Impoundments as Affected by Aeration." _Wat. Poll. Con. Fed._, 41, 5, Part 2, R174-R188 (1969).

Nottingham, P.M. and Hungate, R.E., "Methanogenic Fermentation of Benzoate." _Jour. of Bact._, 98, 3, 1170-1172 (1969).

Novelli, G.D. and Zobell, C.E., "Assimilation of Petroleum Hydrocarbons by Sulfate-Reducing Bacteria." _Jour. of Bact._, 47, 447 (1944).

Oshima, T., "On the Anaerobic Metabolism of Aromatic Compounds in the Presence of Nitrate by Soil Microorganisms." _Z. Allg. Mikrobiol._, 5, 386-394 (1965).

Owen, W.F., Stuckey, D.C., Healy, J.B., Young, L.Y., and McCarty, P.L., "Bioassay for Monitoring Biochemical Methane Potential and Anaerobic Toxicity." _Water Research_, 13, 485-492 (1979).

Painter, H.A., "Biodegradability." _Proc. of Royal Soc. London_, Ser. B, 185 (1079), 149-158 (1974).

Payne, W.J., "Assays for Biodegradability Essential to Unrestricted Usage of Organic Compounds." _Bio-Science_, 20, 15, Aug. 1 (1970).

Pretorius, W.A., "The Effect of Formate on the Growth of Acetate Utilizing Methanogenic Bacteria." _Water Research_, 6, 1213-1217 (1972).

Procter, M.H. and Scher, S., "Decomposition of Benzoate by a Photosynthetic Bacterium." _Soc. Biochem. Jour._, 76, 33P (1960).

Rosenfeld, W.D., "Anaerobic Oxidation of Hydrocarbons by Sulfate-Reducing Bacteria." _Jour. of Bact._, 54, 664 (1947).

Sachs, E.F., Jennett, J.C., and Rand, M.C., "Anaerobic Treatment of Synthesized Organic Chemical Pharmaceutical Wastes." Paper presented at the 33rd Purdue Ind. Waste Conf. May 9-11, 1978 at Purdue Univ., West LaFayette, Indiana.

Saldick, J., "Biodegradation of Cyanuric Acid." Applied Microbiology, 28, 6, 1004-1008 (1974).

SCS Engineers, "Selected Biodegradation Techniques for Treatment and/or Ultimate Disposal of Organic Materials." Technical Completion Report, EPA-600/2-79-006.

Sethunathan, N., Bautista, E.M., and Yoshida, T., "Degradation of Benzene Hexachloride by a Soil Bacterium." Cand. Jour. of Microbiol., 15, 12, 1349-1354 (1969).

Shelton, T.B. and Hunter, J.V., "Anaerobic Decomposition of Oil in Bottom Sediments." Jour. Wat. Poll. Con. Fed., 47, 9, 2257 (1975).

Soap and Detergent Association, "A Procedure and Standards for the Determination of the Biodegradability of Alkyl Benzene Sulfonate and Linear Alkylate Sulfonate." Jour. Am. Oil Chem. Soc., 42, Nov (1965).

Standard Methods Committee-Subcommittee on Biodegradability, "Required Characteristics and Measurement of Biodegradability." J. Wat. Poll. Con. Fed., 39(7), 1232-1235 (1967).

Stanier, R.Y., Sleeper, B.P., Tsuchida, M., and McDonald, D.C., Jour. Bact., 59, 137 (1950).

Stanier, R.Y., Jour. Bact., 59, 527 (1950).

Stanier, R.Y., Jour. Bact., 55, 477 (1948).

Stenerson, J.H.V., "DDT-Metabolism in Resistant and Susceptible Stable Flies and in Bacteria." Nature, 207, 660 (1965).

Stennett, G.V. and Eden, G.E., "Assessment of Biodegradability of Synthetic Surfactants by Tests Simulating Sewage Treatment." Water Research, 5, 601-609 (1971).

Suidan, M.T., Cross, W.H., Fong, M., and Calvert, J.W., "Anaerobic Carbon Filter for Degradation of Phenols." Jour. of Environ. Engr. Div., Proc. Am. Soc. Civil Engr., Vol. 107, No. EE3, June (1981).

Suidan, M.T., Cross, W.H., Fong, M., Calvert, J.W. and Khan, K.A., "Anaerobic Activated Carbon Filters for the Removal of Refractory and Toxic Organic Compounds in Wastewater." Technical Completion Report, ERC-08-79 (1979).

Sweeney, W.A. and Foote, J.K., "A Rapid, Accurate, Test for Surfactant Aerobic Biodegradability." J. Wat. Poll. Con. Fed., 36, 1 (1964).

Tarvin, D. and Buswell, A.M., "The Methane Fermentation of Organic Acids and Carbohydrates." Jour. Amer. Chem. Soc., 56, 1751-1755 (1934).

Taylor, B.F., Campbell, W.L., and Chinoy, I., "Anaerobic Degradation of the Benzene Nucleus by a Faculative Anaerobic Microorganism." Jour. of Bact., 102, 2, 430-437 (1970).

Umbreit, W.W., Burris, R.H., and Stauffer, J.F., Manometric Techniques. (Burgess Publishing Company, Minneapolis, 1957).

Vath, C.A., "Biodegradation of Nonionics, Part I." Soap and Chem. Spec., Feb. (1964).

Vath, C.A., "Biodegradation of Nonionics, Part II." Soap and Chem. Spec., March (1964).

Vicacro, J.P. and Ambye, E.L., "Anaerobic Biodegradability of Carboxymethyloxsuccinate, a Detergent Builder." Jour. of the Amer. Oil Chem. Soc., 54, Jan. (1977).

Voets, J.P., Pipyn, P., Van Lancker, P., and Verstraete, W., "Degradation of Microbicides Under Different Environmental Conditions." J. Appl. Bact., 40, 67-72 (1976).

Weaver, P.J. and Coughlin, F.J., "Measurement of Biodegradability." Jour. of Amer. Oil Chem. Soc., 41, Nov. (1964).

Whittle, P.J., Lunt, D.O., and Evans, W.C., Anaerobic Photometabolism of Aromatic Compounds by Rhodopseudomonas sp." Biochem. Soc. Trans., 4 (1976).

Whittle, P.J., Ph.D. Thesis, Univ. of Wales, Bangor (1975).

Widdus, R. and Ault, C.R., "Progress in Research Related to Genetic Engineering and Life Synthesis." International Review of Cytology, Vol. 38, 1974 (Academic Press).

Williams, R.J. and Evans, W.C., "The Metabolism of Benzoate by *Moraxella* species through Anaerobic Nitrate Respiration." Biochem. J., 148, 1-10 (1975).

Williams, R.J. and Evans, W.C., "Anaerobic Metabolism of Aromatic Substrates by Certain Microorganisms." Biochem. Soc. Trans., Vol. 1 (1973).

Supplemental References
Sugar Manufacturing Wastes

Basu, A.K. and Leclerc, E., "Mesophilic Digestion of Beet Molasses Distillery Wastewater." In "Advances in Wat. Poll. Research." (1972): Chem. Abs., 80, 99891c (1974).

Candelario, R.M., *et al.*, "The Treatment of Liquid Wastes from the Cane Sugar Industry in Puerto Rico." U.S. Natl. Tech. Inform. Serv., Springfield, Va., W. 74-12865 (1974).

Heertjes, P.M. and Van Der Meer, R.R., "Dynamics of Liquid Flow in an Up-Flow Reactor Used for Anaerobic Treatment of Wastewater." Biotechnol. Bioeng., 20, 1577 (1978).

Parthasarathy, T., Thangadurai, D., and Ahmed, N., "Anaerobic Stabilization of Distillery Waste from Cane Sugar Industry." Environ. Health (India) 9, 110 (1967); Water Poll. Abs. (Brit.) 42, 7, 320 (1969).

Pette, K.C., De Vletter, R., Wind, E., and Van Gils, W., "Development and Operation of a Full Scale Anaerobic Plant Treating Sugar Beet Wastewater (in Dutch) Final Report Min. Volksgezondheid en Milieuhygiene, The Hage (1979).

Radhakrishnan, I., De, S.B., and Nath, B., "Evaluation of the Loading Parameters for Anaerobic Digestion of Cane Molasses Distillery Wastes." J. Wat. Poll. Con. Fed., 41, R431 (1969).

Rao, B., Tech Annu. - Indian Assoc. Water Poll. Control, 6-7, 83-90 (1980).

Shakhnovich, B.M. and Sizov, K., "New Process of Sugar Refinery Wastewater Purification." Sakh. Prom. (USSR), 6, 71 (1973); Chem. Abs., 79, 107926q (1973).

Sinha, S.N. and Thakur, B., "Anaerobic Digestion of Cane Sugar Wastes." Environ. Health (India), 9, 118 (1967); Water. Poll. Abs., (Brit.), 42, 2, 320 (1969).

Vitkovskaya, V.A. et. al., "Biochemical Purification of Concentrated Wastewaters from Distilleries Processing Sugarbeet Molasses." Chem. Abs., 79, 60, 4547b (1973).

Brewery Wastes

Basu, A.K. and Leclerc, E., "Comparitive Studies on Treatment of Beet Molasses Distillery Waste by Thermophilic and Mesophilic Digestion." Water Res., (G.B.), 9, 1, 103 (1975).

Bosch, F.M., Roels, R., and Bienvenu, J.M., "Treatment of Brewery Liquid Wastes by Anaerobic Digestion." Trib. CEBEDEAU (Belgium), 21, 295 and 308 (1968); Chem. Abs. 70, 18, 239 (1969).

Halbert, E.J. and Barnes, C.S., Proc. Conv. - Inst. Brew. 17, 219-25 (1980).

Hiatt, W.C., et al., "Anaerobic Digestion of Rum Distillery Wastes." Proc. 28th Ind. Waste Conf., Purdue Univ. (1973).

Kormi, I., "Energy Recovery from Brewery By-Products Through Anaerobic Decomposition." Ph.D. Dissertation, University of Pennsylvania (1976); Diss. Abs. Int., 37, 1805-B (1976).

Koshel, M.I. et al., "Study of the Two-Stage Fermenting of Molasses Distillation Slops by Methane-Forming Bacteria." Mikrobiol. Zh. (USSR), 38, 4443 (1976); Chem. Abs., 85, 121723 (1976).

Lovon, C.R. and Foree, E.G., "The Anaerobic Filter for the Treatment of Brewery Press Liquor Waste." Proc. 26th Ind. Waste Conf., Purdue Univ. (1971).

Nikitin, G.A. and Pavlychenko, L.V., "A Study of the Fermentation of Distillery Molasses Waste by a Mixed Culture of Methane Producing Bacteria." Appl. Biochem. Microbiol., 3, 1 (1968).

Pettet, A.E.J., Tomlinson, T.G., and Hemens, J., "The Treatment of Strong Organic Wastes by Anaerobic Digestion." A Paper presented at a meeting of the Institution of Public Health Engineers at the Conf. Room, Hotel Majestic Harrogate 1959.

Schroeder, E.D., "Pilot Scale Treatment of Wine Stillage." Environ. Protective Technol., Ser., EPA-660/2-75-002 (Feb. 1975).

Animal Wastes

"Animal Manures Into Energy." Energy Digest, 24 (October 1978).

Baines, S., "Anaerobic Treatment of Farm Wastes." Farm Wastes Symp. (G.B.), Newcastle Upon Tyne (1970).

Bell, C., "Anaerobic Digestion of Dairy Farm Slurry." Effl. and Water Trt. Jour. (G.B.), 13, 4, 232 (1973).

Booram, C.V. et al., "Methane Generation from Livestock Wastes in Northern Georgia." Paper Presented at Winter Meeting ASAE. Chicago, Ill. (Dec. 15-18, 1975).

Chen, Y.R. et al., "Methane Production from Agricultural Residues." A Short Review. Ind. Eng. Chem. Prod. Res. Div., 19, 471 (1980).

Chen, Y.R. et al., "Effect of Temperature on Methane Fermentation Kinetics of Beef-Cattle Manure." Biotechnol. Bioeng. Symp., 10, 325 (1980).

Coe, W.B. and Turk, M., "Processing Animal Waste by Anaerobic Fermentation." Symp. Process. Agric. Munic. Waste 29 (1973); *Chem. Abs.*, 81, 158374g (1974).

Converse, J.C. *et al.*, "Dairy Manure Degradation Under Mesophilic and Thermophilic Temperatures." Paper Presented at Winter Meeting ASAE, Chicago, Ill. (Dec. 15-18, 1975).

Fischer, J.R. *et al.*, "Design and Operation of a Farm Anaerobic Digestor for Swine Manure." *Trans. ASAE*, 22, 1129 (1979).

Fischer, J.R. *et al.*, "Anaerobic Digestion in Swine Wastes." In "Energy, Agriculture and Waste Management." W.J. Jewell (Ed.), Ann Arbor Science Publishers, Ann Arbor, Mich. (1975).

"Fuel Gases Produced from Animal-Feed Wastes?" *Chem. Eng.*, 78, 21, 80 (1971).

Gaddy, J.L. *et al.*, "Kinetics and Economics of Anaerobic Digestion of Animal Wastes." *Water, Air, and Soil Pollution*, 3, 161 (1974).

Gerrish, J.B. *et al.*, "Enhanced Treatment of Livestock Wastewater." *Jour. Agric. Engr. Res.*, 20, 1, 13 (1975).

Giesler, F.J., "Is Gas From Manure Practical to Produce and Use." *Hoard's Dairyman*, 120, 1, 40 (1975).

Goodrich, P.R. and Moore, J.A., "Recovery of Energy from Farm Solid Wastes and Timber Production Residues." Final Report, March 7-September 30, 1975." U.S. Energy Res. Devel. Admin. NP-20954 (1975).

Hashimoto, A.G. *et al.*, "Thermophilic Anaerobic Fermentation of Beef Cattle Residue." Report #TID-29414, available from NTIS (1977); *Chem. Abs.*, 91, 1602605 (1979).

Hassan, A.E., *et al.*, "Energy Recovery and Feed Production from Poultry Wastes." In "Energy Agriculture and Waste Management." J.W. Jewell (Ed.), Ann Arbor Science Publishers, Ann Arbor, Mich. (1975).

Hassan, H.M. _et al_., "Characterization of Methane Production from Poultry Manure." Proc. 3rd International Symposium on Livestock Wastes, Univ. Of Illinois, Urbana-Champaign, Illinois (April 21-24, 1975).

Hawkes, D. _et al_., "The Application of Anaerobic Digestion to Producing Methane Gas and Fertilizer from Farm Wastes." _Process Biochem_., (G.B.), 12, 2, 32 (1976).

Hein,, M.E. _et al_., "Anaerobic Digestion of Beef Manure and Corn Stover." Paper Presented at Winter Meeting ASAE, Chicago, Ill. (Dec. 15-18, 1975).

Hobson, P.N. and McDonald, I., "Methane Production from Acids in Piggery Waste Digesters." _Jour. Chem. Tech. Biotechnol_., 30, 405 (1980).

Hobson, P.N. and Shaw, B.G., "The Anaerobic Digesestion of Waste from an Intensive Pig Unit." _Water Res_., (G.B.), 7, 437 (1973).

Ianotti, E.L. _et al_., "Changes in Swine Manure During Anaerobic Digestion." _Dev. Ind. Microbiol_., 20, 519 (1979); _Chem. Abs_., 92, 25451 K (1980).

Ifeadi, C.N. and Brown, J.B., Jr., "Technologies Suitable for Recovery of Energy from Livestock Manure." In "Energy, Agriculture and Waste Management," W.J. Jewell (Ed.), Ann Arbor Science Publishers, Ann Arbor, Mich. (1975).

Kroeker, E.J. _et al_., "Cold Weather Energy Recovery from Anaerobic Digestion of Swine Manure." In "Energy, Agriculture and Waste Management," W.J. Jewell (Ed.), Ann Arbor Science Publishers, Ann Arbor Mich. (1975).

Lai, C.L. _et al_., "A Research Report on the Treatment of Industrial and Agricultural Wastes." _Taiwan Sugar_ (Taiwan) 21, 2, 37 (1974).

Lapp, H.M. _et al_., "Start-Up of Pilot Scale Swine Manure Digesters for Methane Production." Proc. 3rd International Symposium on Livestock Wastes, Univ. of Illinois, Urbana-Champaign, Illinois (April 21-24, 1975).

Lee, Y.B. and Park, C.S., "Production of Methane as a Fuel from Feces of Swine, Fowl and Cattle." Chem. Abs., 76, 22, 131196k (1972).

Loehr, R.C. et al., "Development and Demonstration of Nutrients Removal from Animal Wastes." Environ. Protection Technol. Ser., EPA-R2-73-095 (Jan. 1973); Poll. Abs., 5, 4, 74-03321 (1974).

Maier, W.J. and Fredrickson, A.G., "Recovery of Energy From Farm Solid Wastes and Timber Production Residues." Final Report, March 7-September 30, 1975, U.S. Energy Res. Devel. Admin. NP-20954 (1975).

Manges, H.L. et al., "Treatment and Ultimate Disposal of Cattle Feedlot Wastes." Environ. Protection Technol. Ser., EPA-660/2-75-013 (June 1975).

Meckert, G.W., "Commercial SNG Production from Feedlot Wastes." Symposium Papers: Energy from Biomass and Wastes, Washington, D.C., Institute of Gas Technology, 431 (1978).

Morris, G.R., et al., "Anaerobic Fermentation of Animal Wastes: A Kinetic Design Evaluation." Proc. 32nd Ind. Waste Conf., Purdue Univ. (1977).

Morris, G.R. et al., "Alternative Animal Waste Anaerobic Fermentation Designs and Their Costs." In "Energy, Agriculture and Waste Management," W.J. Jewell (Ed.), Ann Arbor Science Publishers, Ann Arbor, Mich. (1975).

Neel, J.M. and Zeigler, L., "Processing of Animal Wastes by Anaerobic Fermentation." Paper Presented at the 69th Annual AIChE Meeting, Windsor Locks, Conn., Nov.-Dec., 1976.

Ostrander, C.E., "Techniques That are Solving Pollution Problems for Poultrymen." Proc. 3rd International Symposium on Livestock Waste, Univ. of Illinois, Urbana-Champaign, Illinois (April 21-24, 1975).

Patel, J.B. and Patel, R.B., "Biological Treatment of Poultry Manure Reduces Pollution." Compost. Sci., 12, 5, 18 (1971).

Pratt, M., "That Manure Pit: Is It a Goldmine?" Agricultural Engineering, 56, 6, 19 (1975).

Rengasamy, G. et al., "Anaerobic Treatment of Dairy Wastes." Poll. Abs., 4, 4, 73-4TF-00483 (1973).

Savery, C.W. and Cruzan, D.C., "Methane Recovery From Chicken Manure Digestion." J. Wat. Poll. Con. Fed., 44, 2349 (1972).

Schmid, L.A. and Lipper, R.I., "Swine Waste Digestion Enhancement with Nutrient Separation." U.S. Natl. Tech. Inform. Serv., Springfield, Va., PB-255-021 (1976).

Schmid, L.A. et al., "Separating Nutrients to Enhance Swine-Waste Digestion." Proc. 3rd International Symposium on Livestock Waste, Univ. of Illinois, Urbana-Champaign, Illinois (April 21-24, 1975).

Schmid, L.A. and Lipper, R.I., "Biological Treatment of Beef Animal Wastes." U.S. Natl. Tech. Inform. Ser., Springfield, Va., Govt. Rept. Announcements, 72, 3, 111 (1972).

Schmid, L.A. and Lipper, R.I., "Swine Wastes, Characterization and Anerobic Digestion." Cornell Univ. Conf. Agr. Waste Management. Ithaca, N.Y., 50 (1969); Water Poll. Abs., 44, 6, 1169 (1971).

Shelef, G. et al., "High-Rate Thermophilic Anaerobic Digestion of Agricultural Wastes." Biotechnol. Bioeng. Symp., 10, 341 (1980).

Shih, J.C.H. and Huang, J.J.H., "A Laboratory Study of Methane Production from Broiler-Chicken Litter." Biotechnol. Bioeng. Symp., 10, 317 (1980).

Smith, K., "Operation of an Anaerobic Digester at the Washington State Dairy Farm." Symposium Papers: Energy From Biomass and Wastes, Washington, D.C., Institute of Gas Technology, 403 (1978).

Smith, R.E., et al., "Two-Phase Anaerobic Digestion of Poultry Waste." Paper presented at Winter Meeting ASAE, Chicago, Ill. (Dec. 15-18, 1975).

Turnacliff, W. et al., "Anaerobic Digestion of Poultry Wastes and Algae Production from Digester Effluent." Report #NP-23598 available from Bio-Gas, Colo., Inc., Res. Div., Arvada, Colo. (1978); Chem. Abs., 91, 198307A (1979).

Van Faassen, H.G. and Van Dijik, H., "Oxidation-Nitrification and Dent··ification of Veal Calf Manure." Proc. 3rd I ernational Symposium on Livestock Wastes, Univ. of Illinois, Urbana-Champaign, Illinois (April 21-24, 1975).

Van Velsen, A.F.M., "Anaerobic Digestion of Wet Piggery Waste." Comm. Eur. Communities, EUR6249, 476 (1979); Chem. Abs., 92, 25437K (1980).

Van Velsen, A.F.M., "Anaerobic Digestion of Piggery Waste. II. Start-Up." Neth. Jour. Agric. Sci., (Neth.), 27, 142 (1979).

Van Velsen, A.F.M., "Anaerobic Digestion of Piggery Waste III. Influence of Temperature." Neth. Jour. Agric. Sci., (Neth.), 27, 255 (1979).

Wallingford, G.W. et al., "Denitrification in Soil Treated with Beef-Feedlot Manure." Commun. In Soil Science and Plant Analysis, 6, 2, 147 (1975).

Wolf, R., "Methane Generation on a Farm." Compost. Sci., 16, 5, 28 (1975).

Wujcik, W.J. and Jewell, W.J., "Dry Anaerobic Fermentation." Biotechnol. Bioeng. Symp., 10, 43 (1980).

Yokoyama, M., "Production of Methane Gas from Chicken Excrement." Chem. Abs., 81, 158466p (1974).

Food Processing

Aulenbach, D.B. and Hallock, R.V., "Treatment of Milk Truck Washing Waste by Anaerobic Digestion." Proc. 27th Ind. Waste Conf., Purdue Univ. (1972).

Bizek, E. and Filla, O., "Biochemical Purification of Waste Waters from the Washing of Greasy Wool." Chem. Abs., 79, 23401b (1973).

Borchardt, J.A. and Pohland, F.G., "Anaerobic Treatment of Alewife Processing Wastes." J. Wat. Poll. Con. Fed., 42, 2060 (1970).

Chadwick, T.H. and Schroeder, E.D., "Characterization and Treatability of Pomace Stillage." J. Wat. Poll. Con. Fed., 45, 1978 (1973).

Hakannson, H. et al., "Anaerobic Treatment of Whey and Whey Permeate in Submerged Filters." Inst. Vatten-Luftvardsforsl. Publ. B, 406 (1977).

Hudson, J.W. and Pohland, F.G., "Anaerobic Packed Column Treatment of Shellfish Processing Wastewater." Proc. 33rd Ind. Waste Conf., Purdue Univ. (1978).

Kissam, A. et al., "Preliminary Evaluation of Anerobic Sludge Digestion for the Tuna Processing Industry." Proceedings 8th National Symp. Food Process. Wastes, EPA-600/2-77-184, 155 (1977); Chem. Abst., 92, 10751J (1980).

Koehler, R., "Anaerobic Decomposition of Yeast Factory Waste Water." Wasser, Luft Betr. (Ger.), 17, 10, 342 (1973); Chem. Abs., 80, 63588v (1974).

Lane, A.C., "Methane from Anaerobic Digestion of Fruit and Vegetable Processing Wastes." Jour. Food Tech. Australia. (Austr.), 5, 201 (1979).

McCarty, P.L. et al., "Treatability of Oily Wastewaters from Food Processing and Soap Manufacture." Proc. 27th Ind. Waste. Conf., Purdue Univ. (1972).

McMorrow, N.C. *et al*., "Anaerobic Digestion of Pineapple Mill Waste." Proc. 25th Ind. Waste Conf., Purdue Univ. (1970).

McNary, R.R., Wolford, R.W., and Dougherty, M.H., "Experimental Treatment of Citrus Waste Water." Proc. 8th Ind. Waste Conf., Purdue Univ. (1952).

Olivier, G.E. and Dunstan, G.H., "Anaerobic Digestion of Pea-Blancher Waste." *Sewage and Industrial Wastes*, 27, 10, 1171 (1955).

Taylor, D.W. and Burn, R.J., "Full Scale Anaerobic Filter Treatment on Wheat Starch Plant Wastes." Amer. Inst. Chem. Engr. Symp. Ser., 69, 129, 30 (1973); *Chem. Abs*., 79, 34816S (1973).

University of Newcastle, "Symp. Rept. Treatment of Wastes from the Food and Drink Industry." *Process Biochem*. (G.B.), 9, 2, 26 (1974).

Van Den Berg, L. and Lentz, C.P., "Food Processing Waste Treatment by Anaerobic Digestion." Proc. 32nd Ind. Waste Conf., Purdue Univ. (1977).

Van Den Berg, L. and Lentz, C.P., "Anaerobic Digestion of Pear Waste: Factors Affecting Performance." Proc. 27th Ind. Waste Conf., Purdue Univ. (1972).

Van Den Berg, L. and Lentz, C.P., "Anaerobic Digestion of Pear Waste: Laboratory Equipment Design and Preliminary Results." *Can. Inst. Food Tech. Jour*., 4, 4, 159 (1971).

CHAPTER 3

TOXICITY EFFECTS IN ANAEROBIC PROCESSES

INTRODUCTION

Adequate knowledge of the toxicity of relevant toxins and inhibitors in any biological process is essential for an optimal application of the process. Engineers designing facilities for removal of organics from industrial wastewaters often will choose conventional aerobic treatment where, in many cases, anaerobic treatment would be satisfactory. When considering the many advantages of anaerobic treatment, the question has to be asked why the application of methane fermentation is so limited in this country in the treatment of industrial wastewaters.

Much of the reluctance to use anaerobic processes stems from the belief that the methane fermentation systems cannot tolerate chronic or slug doses of toxic substances found in industrial wastewaters. There are many well documented cases in the literature where the presence of toxicants may have caused inhibition which eventually lead to failure of the process, particularly in the case of anaerobic sludge digestion. However, it should not be assumed that methanogens are more sensitive to toxicants than facultative organisms.

During the past 15 years there have been many documented studies concerning possible inhibitory effects on the anaerobic digestion process from from compounds commonly found in these systems. In many instances in which municipal digester performance had decreased, the source of the upset could be traced to the presence of a compound that was inhibitory to the microorganisms involved in the digestion process. These compounds originate from waste streams generated by different industrial processes, some of which are listed in Table 3.1.

Table 3.1 Point Source Categories [Keith and Telliard, 1979]

Category
Timber products processing
Steam electric power plants
Leather tanning and finishing
Iron and steel manufacturing
Petroleum refining
Inorganic chemicals manufacturing
Textile mills
Organic chemicals manufacturing
Nonferrous metals manufacturing
Paving and roofing materials
Plant and ink formulation and printing
Paint and ink
Printing and publishing
Soap and detergent manufacturing
Auto and other laundries
Plastic and synthetic materials manufacturing
Pulp and paperboard mills and converted paper products
Rubber processing
Miscellaneous chemicals
Adhesives
Gum and wood chemicals
Pesticides
Pharmaceuticals
Explosives manufacturing
Machinery and mechanical products manufacturing
Aluminum forming
Battery manufacturing
Coil coating
Copper forming
Foundries
Plastics processing
Porcelain enamel
Mechanical products
Electrical and electronic components
Electroplating
Ore mining and dressing
Coal mining

These waste streams, which may or may not receive pretreatment at the industrial plant site, are released into municipal sewers where they are transported to the local municipal plant for final treatment. At the plant, these compounds eventually wind up in the sludge being digested. Due to the large number of compounds generated and the geographical specifity of the industries that produce them, the available literature contains numerous, but isolated, studies.

Through a series of articles printed nearly 20 years ago, McCarty (1964) was able to summarize the available literature on anaerobic waste treatment. Among the topics covered, a summarization on toxic materials and their control was included. At the time the report was written, the only anaerobic process used commercially to any great extent was the conventional digestion system, and its primary application was to stabilize sewage sludge solids. Toxicity problems reported up to that period pertained solely to the conventional process. As was the case with the information on the anaerobic digestion processes, the available information on the type and concentration of compounds that produced inhibitory and/or toxic effects on the anaerobic organisms involved was also limited. The compounds focused on up to 1964 were mainly limited to the heavy metals, alkali and alkaline-earth salts, ammonia, sulfides and a few organic compounds.

In the past 10 years there have been several significant advances in the field of anaerobic waste treatment, including the development of several new processes (see Chapter 1) as well as new techniques for measuring and predicting inhibition in an anaerobic system. Also, during this period, a large number of compounds have come into focus as being suspect for being toxic to the microorganisms involved in carrying out the mechanisms of anaerobic fermentation.

In Chapter 1, the systematic development of several different types of anaerobic treatment processes were described in detail. In similar fashion the purpose of Chapter 3 will be to describe and evaluate the advances made in the area of toxicity effects in anaerobic waste treatment. The following areas will be described and evaluated in the chapter:

1. Inhibition, Toxicity, Antagonism and Synergism.

2. Anaerobic Tests Used for Measuring Toxicity.

3. Toxicity Models.

4. Listing of Compounds known to be Toxic to Anaerobic Systems.

INHIBITION, TOXICITY, ANTAGONISM, AND SYNERGISM

Before any tests and models can be described and properly understood, certain parameters must first be defined. Perhaps the best way to explain the difference among stimulation, inhibition, and toxicity and, at the same time, illustrate the effect of a compound on a biological system is to present Figure 3.1. Figure 3.1 is a general plot of the rate of biological reaction versus increasing concentration of compound as described by McCarty (1964) and is similar to the graph presented by Kugelman and Chin (1971). The figure is representative of what happens when certain organic and in organic compounds are added to biological systems.

Point A in the figure represents the rate at which biological reactions occur prior to the addition of the example compound. Addition of the compound in amounts that produce values below the optimum concentration usually result in a stimulation of biological activity as represented in an increase in the reaction rate of the organisms present. Point C is considered the optimum concentration value since the organisms are at their highest rate. Any additional amount of compound added that results in concentrations greater than the optimum but less than the crossover concentration will produce a decrease in the biological activity of the system. As was the case with Region B where the organisms were being stimulated, the same organisms in Region D were now considered to be inhibited by the compound added at those concentrations. If the compound is present in concentrations greater than the crossover concentration, the biological activity of the system is now at a value less than when the compound was absent from the system. When concentrations are

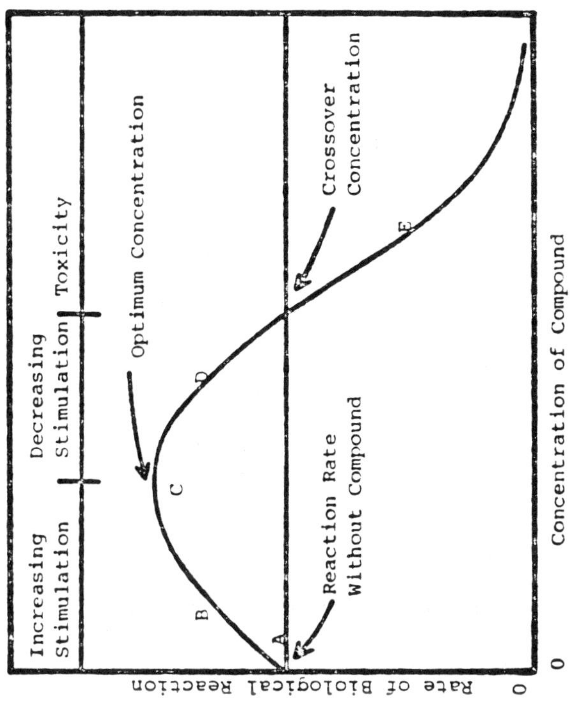

Fig. 3.1 General Effect of Compound on Biological Reaction (Mc Carty, 1964)

into Region E, the compound (at those concentrations) is considered toxic to the organisms present, and, if added to produce high enough concentrations, biological activity will essentially cease.

In general, toxicity effects are highly dependent upon concentration and type of material present in the system. When a compound is added to a biological system, the concentration level that is maintained dictates whether or not the biological activity within the system is stimulated, inhibited or ceases altogether. Also, toxicity effects are dependent upon the type of material added to the system, since some compounds are more toxic than others. In terms of Figure 3.1, the compound considered more toxic will inhibit the microorganisms present at lower concentrations than the less toxic materials.

When reviewing toxicity data reported in literature, special attention should be paid to the environmental conditions under which toxic levels are observed. There are several known mechanisms that have the ability to increase or decrease the magnitude of toxicity experienced in a biological system caused by the presence of a certain material. If the researcher is unaware that any of these mechanisms is functioning in the systems at the time the observation is made, erroneous conclusions about the toxic effects on a biological system from a particular substance are likely to be made. The above-mentioned mechanisms are listed as follows and subsequently will be reviewed:

1. Antagonism and Synergism
2. Complex Formation
3. Microbial Growth
4. Acclimation
5. Volatilization

Antagonism and Synergism

Antagonism is the condition where the toxicity effects caused by the original material is reduced due to the presence of a second. Synergism, on the other hand, is the condition where the toxicity effects caused by the original material is increased by a second material in the system. As was the case with toxicity, both synergism and antagonism are dependent upon concentration [Kugelman and Chin, 1971]. Perhaps the best way to illustrate the working of antagonistic and synergistic relationships is to review Figure 3.2. Figure 3.2 was originally developed by Kugelman and McCarty (1965) and is based on an arbitrary dual system of compounds A and B. Compound A is introduced into a system at concentrations that produce toxic conditions, the reaction rates of the organisms at this point are measured and used as a control value. Compound B is now introduced in increasing amounts in order to measure the pattern of the antagonistic or synergistic relationship between the two compounds.

As compound B is added, if the effects are antagonistic the reaction rate of the organisms involved will increase. On the other hand if the addition of compound B reduces the reaction rate of the organisms further, then the effects are synergistic. In both cases the reaction rate is compared against the control value and plotted as a ratio, as show in the figure.

For the antagonistic response pattern, as the concentration of compound B increases (even at low concentrations of B) and the concentration of compound A stays constant, the toxic effect produced by compound A decreases as shown by the increases in the reaction rate of the organisms involved. After increasing the concentration of B still further, the antagonistic effects reach a peak value followed by a decrease in the reaction rates. This decrease continues as the concentration of B increases till the cross-over concentration is obtained. The cross-over concentration represents the concentration at which the reaction rate in the system containing both A and B equal the reaction rate in a system containing only A. Any further increase in B will decrease the reaction rates even lower.

In the synergistic system, low concentrations of B have little effect on the reaction rates of the organisms in the system. As the concentration of B is continually increased, the reaction rate steadily declines.

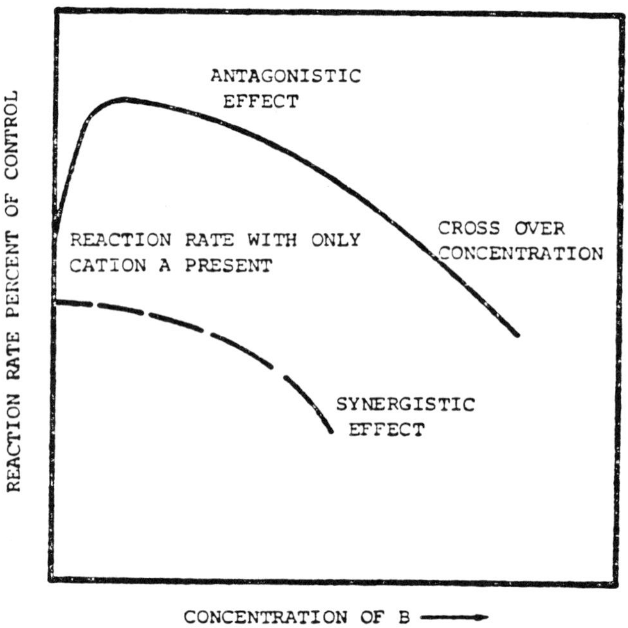

Figure 3.2 Typical Antagonistic and Synergistic Relationships in a Two Cation System With the Concentration of Cation A Held at a Toxic Level (Kugelman and McCarty, 1965)

The case described above represented a dual system of antagonism since only one material was present to antagonize the substance causing toxicity. There are cases where combinations of compounds act together to antagonize the toxic material. In this type of system, the first material is described as the primary antagonist while the second is labeled the secondary antagonist. A primary antagonist is a compound which produces antagonism in a dual system (as described above) while a secondary antagonist is a substance which antagonizes the toxic compound only when the primary antagonist is present [Kugelman and McCarty, 1965].

Even though antagonistic and synergistic relationships have been observed for only a few materials, it should not be assumed that they do not exist for other compounds. The available information dealing with antagonism and synergism is centered around the light metal cations such as potassium, magnesium, calcium and sodium [Kugelman and McCarty, 1965; Kugelman and Chin, 1971]. From these studies it was found that only small concentrations of an antagonistic or synergistic compound were required to invoke a change in the reaction rate of microorganisms. Also, combinations of materials were found to produce either antagonistic or synergistic effects, but not both.

Hovious, et al. (1973) conducted a study measuring the effects (inhibitory, toxic, etc.) that organic compounds commonly found in the wastewaters of the petrochemical industry had on the mechanisms of the anaerobic digestion process. One particular aspect of the study was concerned with the effect a combination of these compounds had on the anaerobic digestion process. Initially, slowly increasing but equal quantities of four known inhibitors (phenol, acrylonitrile, ethyl acrylate and formaldelhyde) were injected into a synthetic waste being fed to an anaerobic filter reactor. They were able to conclude that the compounds act synergistically, causing the reactor to be inhibited at concentration levels below those that caused inhibition when each of the compounds had acted alone.

To help understand the true nature of toxic effects that compounds exert on the mechanisms of anaerobic fermentation, synergistic and antagonistic relationships between compounds other than the metal cation group should be examined. However, at this point, the information available is very limited.

Complex Formation

Complex formation, like antagonistic relationships between compounds is a mechanism that reduces the magnitude of toxicity experienced by the system. By forming a complex, the compound is not in the proper chemical form to gain access into the cell. By not gaining access into the cell, the metabolic activity of the organism is unchanged.

In order to predict the magnitude that will occur due to the materials present, complex formation like antagonistic and synergistic relationships should be taken into account. In some instances due to the materials present, complex formation can be easily predicted according to stoichiometric relationships. However, in some cases it is more difficult to predict complex formation since some compounds will not bind until they have been transformed into a suitable condition by microbial activity [Kugelman and Chin, 1971].

Perhaps the best example of complex formation in relation to anaerobic waste treatment is the complexation of the heavy metals and sulfides. Heavy metals have been blamed for many anaerobic treatment failures [Regan and Peters, 1970; Lawrence and McCarty, 1965; Rudgal, 1946]. Low but soluble concentrations of zinc, copper, chromium and nickel salts are toxic in their free ionic form and it is these metals which have been associated with most of the problems of heavy metal toxicity in anaerobic treatment.

In anaerobic digesters that contain heavy metals, the presence or addition of sulfides result in the formation of extremely insoluble metal sulfide salts. The solubility and solubility products of various metal sulfide salts are listed in Table 3.2.

Unless the metals are of unusually high concentrations, there are usually sufficient sulfides present to precipitate the metals and thus place them in a non-toxic form. Sulfides in most cases are biologically produced in the digester by the

Table 3.2 Metal Sulfide Solubility Product Constants (Dickerson, *et al*. 1974)

Heavy Metal	Sulfide Salt	Solubility Product	Solubility (mg/l)
Cadmium	Cds	1.0×10^{-28}	1.0×10^{-9}
Copper	CuS	8.0×10^{-37}	9.0×10^{-14}
Cobalt	CoS	5.0×10^{-22}	2.0×10^{-6}
Iron	FeS	4.0×10^{-19}	6.0×10^{-5}
Lead	PbS	7.0×10^{-29}	2.0×10^{-9}
Nickel	NcS	3.0×10^{-21}	5.0×10^{-6}
Zinc	ZnS	2.5×10^{-22}	2.0×10^{-6}

reduction of sulfates or other sulfur-containing organics or released from the degradation of sulfur containing organic compounds such as proteins [Lawrence and McCarty, 1965]. If heavy metal concentrations are particularly high, and sufficient sulfides are not formed during waste treatment then sodium sulfide or a sulfate salt (which will be reduced to sulfide under anaerobic conditions) can be added.

Another complexation method is a natural process, by which heavy metals are removed and/or reduced by precipitation of the metals as an insoluble carbonate salt. This method gives protection only against certain heavy metals, and only if the pH value of the digester sludge is high enough (pH value greater than or equal to 7.2 for cadmium and greater than or equal to 7.7 for zinc). The best example of this protection is against iron toxicity and is good at all pH values greater than 6.4 which permits any sulfide initially present as ferrous sulfide to be released to precipitate other metals without simultaneously releasing high concentrations of ferrous ions into solution [Mosey, 1976].

Microbial Growth

Microorganisms involved in biological waste treatment processes can be rendered metabolically inactive as a result of reactions occurring between inhibitory compounds present in a system and some essential protein or molecular group of the organism. Another inhibition mechanism involves the adsorption of the toxic compound onto the surface of the cell. A reaction of this type could possibly change the cell's permeability. Inhibition of enzymes is a result of the inhibitory substance and substrate competing for the same attachment sites on the enzyme, destruction of an essential functional group of the enzyme or an attraction of the inhibitory compound for part of the enzyme [Hovious, <u>et al</u>. 1973].

The ability of a compound to inhibit a biological waste treatment process is also a function of the type of growth exhibited by the microorganisms involved. In a system containing fast growing organisms, there is a greater likelihood that inhibition would occur, since the substrate uptake and conversion rate are greater than on a system of slow growers. Also, it would be expected that inhibition effects would be greater in flocculated growth systems than slime or attached growth systems. In flocculated growth systems more cells are exposed to the inhibitor than in the attached growth system, where only the outside layer of cells are exposed to the inhibitor [Hovious, _et al_. 1973].

Acclimation

When an inhibitory compound is introduced into a biological system and if the compound is in high enough concentrations, the result is usually an inactivation of the metabolic activity of the organisms. The phenomenon of acclimation is referred to as the ability of the organism to overcome the adverse effects caused by the inhibitory compound. Rarely, if ever, is all of the toxic compound removed from a system. Therefore, this suggests that the mechanisms of acclimation represent a rearrangement in the metabolic pathways of the organisms instead of a development in a mutation or survival of the fittest [Kugelman and Chin, 1971].

The rate in which an inhibitory compound is introduced into a system can have an effect on the organism's ability to acclimate to the compound. The means by which an inhibitory compound can be introduced into a system is by either slug or gradual dose. Several studies have shown that inhibition caused by a compound is not as severe if the compound is added gradually to the system instead of in a slug dose.

Two studies in particular [McCarty and Kugelman, 1961; Kugelman and McCarty, 1965] were undertaken to determine the nature of salt toxicity in the anaerobic digestion process. Both studies injected laboratory-sized conventional style digesters with varying levels of cations such as calcium, sodium, potassium and magnesium, in order to determine the toxic concentration of each. Both studies were able to conclude that the cations injected on a slug basis were more toxic than when added gradually. By gradually adding the cations, the organisms were able to acclimate to concentrations 2 to 3 times the slug dose concentrations that caused toxicity.

The final work discussed concerning acclimation is the study that was conducted at Drexel University [Parkin, et al. 1980; Yang, et al. 1980; Parkin and Kocher, 1980 a,b]. The main purpose of the study was to evaluate the response of methane fermentation systems to exposure of some 30 organic and metallic pollutants commonly found in industrial wastewaters. In the study, toxic materials were added to digesters operating in different modes of hydraulic flow: batch, semi-continuous and continuous. For the batch and semi-continuous feed studies, a procedure utilizing serum bottles similar in nature to the one developed by Owen, et al. (1979) was used. The continuous studies on the other hand used 1-liter laboratory sized anaerobic upflow filters.

The batch systems were fed only at the beginning of that test during which a given dose of toxic material was also added. Several units for each compound tested was required so that the effects over a wide range of concentrations could be observed. In the semi-continuous studies the serum bottles were fed once a day with the proper amounts of substrate and test compound. Also, at this time, a predetermined amount of solids was removed to provide the system with the desired solids retention time. The toxicant was added to the continuous flow units in two different fashions. In the first method, the toxicant was added continuously into the units along with the substrate while with the second method the toxicant was added for only 1 day periods. For all the tests, the amount of gas produced was used as a measure of metabolic activity and was compared against a control unit (i.e., operated under identical conditions as the others but did not contain the toxic material) to determine the degree of inhibition.

Even though the work has not yet been completed, some results have been published. In particular, results have been published for nickel, ammonia, sulfide and formaldehyde by Parkin, et al. (1980). In general the results obtained were similar and, therefore, nickel will be used as a representative example.

In the batch experiments, unacclimated methanogenic organisms were fed nickel concentrations ranging from 50 to 500 mg/l. A concentration of 200 mg/l of nickel showed 86 percent of the gas production of the control value while a concentration of 300 mg/l of nickel showed only 43 percent of the control value. In the semi-continuous systems, gas production was inhibited by concentrations of 80 to 100 mg/l of nickel. By the end of 10 days, however, all systems recovered and were producing gas values equal to the control. Nickel was added continuously in the feed to an anaerobic filter for over 30 days. The results indicated that once acclimated, 250 mg/l of nickel could be handled by the anaerobes. Using a different filter, 1-day slug additions were made to the filter. The results showed that slug doses of up to 500 mg/l could be handled with only a 25 percent decrease in gas production. After a period of about 9 days the gas production had returned to the value before the compound was introduced into the system.

From these results, Parkin, et al. (1980) were able to conclude that methanogenic bacteria have considerable potential to acclimate to toxic substances found in waste treatment systems. From the results obtained, using the compounds listed above, they found that with acclimation, toxic compounds 2.4 to 12 times those causing inhibition to unacclimated bacteria could be tolerated with no decrease in gas production.

The concentration and exposure time of a toxic compound being introduced into a system plays a significant role in the organism's ability to acclimate to the compound. By introducing a toxic compound gradually instead of in slug doses, it allows time for the organisms to acclimate to the compound. Even though slug doses to biological treatment systems sometimes do occur, the general trend is that the inhibitory material concentration tends to build up gradually in a system instead of suddenly. This phenomenon does allow some acclimation to occur in actual systems.

Volatilization

Volatilization occurs when a gas stream flows through the reactor volume, with some of the toxic material transferred to the gas phase. This allows the material transferred to be removed from the system along with the exiting gas. The volatilzation process can be useful in aiding an anaerobic system to recover from inhibitory/toxic effects caused by organic toxicant additions. Concentration levels of an organic toxicant causing inhibitory/toxic effects may be reduced through the mechanisms of volatilization, to concentrations that do not inhibit the microorganisms within the system.

According to Kirsch, *et al*. (1981), removal by volatilization is a mass transfer phenomenon and the rate of removal of the pollutant will be equal to the product of a mass transfer coefficient and a driving gradient. The equation for the removal of organic compounds is given as follows:

$$r_v = K_v (S_o - S_e) \qquad (3.1)$$

where:

r_v = rate of removal of material by volatilization, mass/volume - time

K_v = overall volumetric mass transfer coefficient for volatilization, 1/time

S_e = concentration of pollutant that would be in the liquid if the liquids were in equilibrium with gas above liquid, mass/volume

S_o = pollutant concentration in the influent to system mass/volume

The driving gradient is represented by $(S_o - S_e)$ and it is the difference between the concentration of the pollutant and the concentration that would be in the liquid if it were in equilibrium with the gas flowing through the reactor. In the case of an anaerobic system, the gas flowing through the system is primarily methane and carbon dioxide generated by the organisms as a by-product of waste stabilization. Once the overall mass transfer coefficient is determined and the steady-state concentration of the pollutant inside the reactor is known, the total amount of pollutant removed by volatilization is equal to r_v times the volume of the reactor.

There have been several studies conducted to determine the concentration levels of certain pollutants that cause toxic effects to the anaerobic digestion process. In several instances, the systems were able to recover and return to normal operation after a suitable period of time. In order to determine the reason for this phenomenon, the studies analyzed the contents of the reactors to determine the fate of the pollutant during the recovery period. Several studies determined that the pollutant was removed by the mechanism of volatilization in such quantities that the pollutant's concentration level (which had caused inhibition to organisms in the reactor) was reduced to the non-inhibitory range. Pearson, _et al_. (1980), while developing inhibition models, had determined that an anaerobic system severely inhibited due to the presence of formaldehyde was able to recover. The major reason for system recovery in this case was attributed to the mechanism of volatilization reducing the formaldehyde concentration to non-inhibitory levels. Similarly, Stuckey, _et al_. (1980) attributed the recovery of a semi-continuous anaerobic digestion reactor (initially inhibited by vinyl acetate) to the removal of the pollutant by volatilization.

ANAEROBIC TESTS USED FOR MEASURING TOXICITY

Most anaerobic digester failures are blamed on the presence of toxic compounds in the system. Due to the complex make-up of most sludges, when digester failure does occur the cause is very difficult to ascertain. Unless there is evidence to suspect a particular material is the cause of the digester failing, chemical analysis alone is useless. In recent years as the popularity of anaerobic waste treatment has increased, so has the interest in determining the effects organic and metallic compounds commonly found in industrial wastewaters have on the anerobic bacteria carrying out the methane fermentation process. As a result, several new tests have been developed for determining whether or not the presence of a particular compound inhibits the bacteria involved in the process. The tests that have been developed and used in various studies to date include:

1. Batch Test

2. Semi-Continuous Test

3. Reversibility Test

4. Continuous Flow Test

The purpose of the tests, with the exception of the reversibility test, is to determine whether or not certain materials are inhibitory to the anaerobic digestion process. To accomplish this, the performance (metabolic activity) of the units containing the material to be evaluated is compared against that of a control unit. A control unit is identical in nature to the other units used in the study, except it does not contain the compound to be examined.

In order to monitor the activity of the microorganisms in anaerobic systems, metabolic parameters such as cell growth, organic reduction, and gas production are measured throughout the span of the experiment and are compared against the activity of the control unit. The majority of the studies to date have used gas production as a means to monitor the metabolic activity of the organisms. Since a large number of concentrations and replicates of each may be evaluated in a study measuring the growth rate of the organisms and/or organic reductions (efficiency in terms of percent removal) would, in many cases, be considerably more time consuming.

Before the anaerobic tests are discussed in detail, certain points must be made clear. Even though certain toxicity assay tests can be used to determine if a compound is biodegradable (see Chapter 2), their main purpose is to determine whether or not the test compound is toxic to the anaerobic organisms and at what concentration levels. The effects on the methanogenic bacteria are usually the primary focus of the studies, since the fermentation of methane from volatile acids is considered to be the rate-limiting step in the process. Also, literature has shown that most anaerobic process failures stem from the inhibition of the methanogenic bacteria, which are apparently most sensitive.

Batch Toxicity Assays

Anaerobic batch toxicity tests are used to determine the effect a test material has on the anaerobic organisms of the digestion process. In biodegradation tests the compound being tested is the sole carbon source for the bacteria. If the organisms are able to metabolize the compound directly, then it is considered biodegradable. A batch toxicity test on the other hand determines the effect a test compound has on organisms directly metabolizing a carbon source other than the test compound. The basic procedure for all anaerobic batch toxicity tests are similar and will be discussed briefly.

It is generally conceded that the methane fermentation step is the rate-limiting step of the anaerobic digestion process [McCarty, 1966]. During the first stage of the anaerobic digestion process, acetic and propionic acids are the most prevalent volatile acid intermediates formed from the fermentation of complex organic materials. Acetic acid is responsible for 70 percent of the methane gas produced in the second stage with only 15 percent from the propionic acid [Jeris and McCarty, 1965; McCarty, _et al_. 1963]. However, it should be pointed out that, to date, only pure cultures of methogens growing on either formate or acetic acid have been isolated.

It is for the above reasons that the anaerobic batch toxicity tests are concerned with the response of those anaerobic bacteria that utilize primarily acetic acid when the test material is present. To begin the experiment, a reactor vessel ranging in volume from 125 ml to 1.5 liters is required. Under anaerobic conditions, everything required to sustain an anaerobic population, i.e. seed organisms, dilution water, carbon source and nutrients is added to the unit at the beginning of the experiment, with no further additions made. This procedure represents a true batch experiment.

The seed organisms are developed from anaerobically digested sludge which, in most cases, is taken from a municipal digester. The digested sludge is placed in a batch laboratory-sized digester and is fed the substrate to be used in the assay. The digester is operated on a semi-continuous basis with the hydraulic and solids retention time ranging from 15 to 50 days. By applying this technique through several retention times, the original population of organisms found in the sludge is removed and replaced by a population of organisms that utilize the applied substrate as its main carbon source. The unit is usually operated at a temperature of 35°C, which is optimum for the mesophilic phase.

According to Jeris and McCarty (1965), anaerobic toxicity is determined as the adverse effect of a substance on the predominant methanogens, i.e. those that convert acetic acid to methane gas. To determine the effect that the added compound has on the methanogens present, the performance of the unit containing the test compound is compared against that of a control unit. The metabolic parameter used to monitor the activity of the organism is total gas (or methane) production. Since a limited amount of substrate is added to the reactor (and only once), a specific and finite quantity of gas is produced over a subsequent time period ranging from 8 to 30 days. The amount of methane gas produced from acetic acid is determined by the following stoichiometric relationship:

Sodium Acetate:

$$CH_3COONa + H_2O \rightarrow NaHCO_3 + CH_4 \qquad (3.2)$$

It should be noted that acetate is added as sodium or potassium salts, and the subsequent reactions produce only methane gas.

To fully test a compound, the substance is added at increasing concentrations to different units, to provide a range of effects increasing from non-inhibitory to severely toxic. To illustrate this point [Parkin, _et al_. 1980], Figure 3.3 shows the effect of the element nickel when added in increasing concentrations (50-500 mg/l) to a seed of methanogenic bacteria. Total gas production over an 8 day period can be compared to the control's. As the dose was increased, not only was the rate of methane gas reduced, but also the total quantity produced. In these tests, several replicates are run for the control and each concentration. Values of gas production shown in Figure 3.3 were the average of the replicates (in this case 6 control units and 3 units for each concentration were run).

The relative amount of inhibition created by that concentration of test material was reflected in the maximum rate ratio [Owen, _et al_. 1979]. The maximum rate ratio, MRR, is defined as follows:

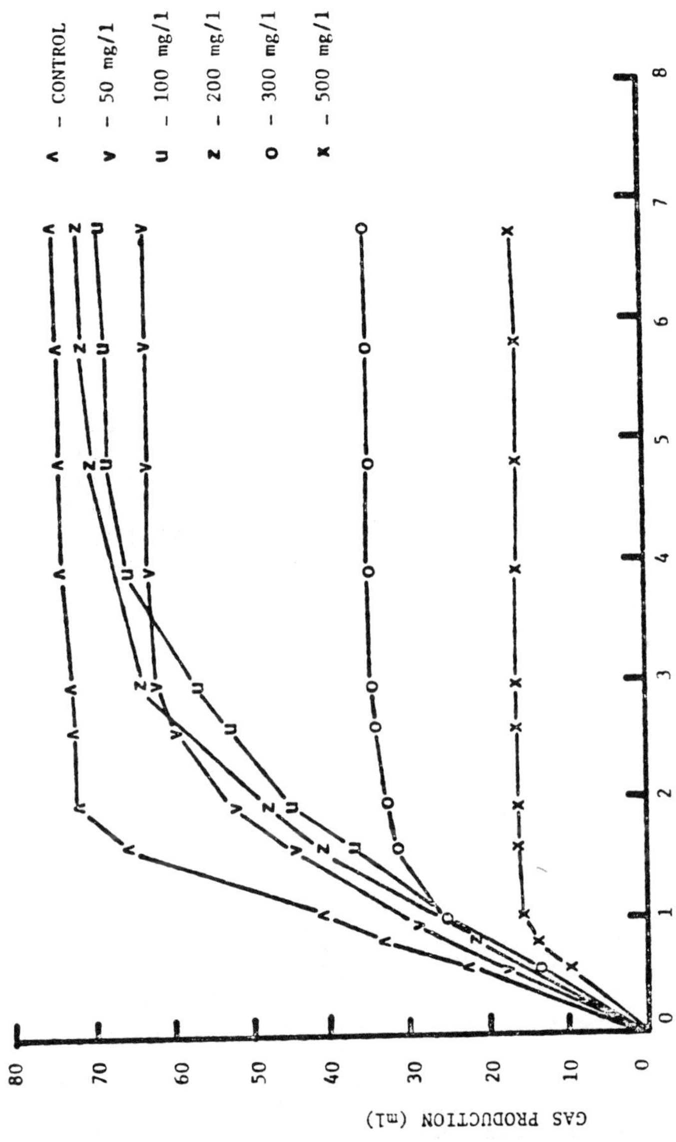

Fig.3.3 Response of Unacclimated Methanogens to Nickel (Parkin, *et al.* 1980)

MRR = [Gas Production With Test Compound Over Specific Time Period]/[Gas Production of Control] (3.3)

The presence of a material can affect the unit´s metabolic activity as shown through the total gas output in one of the following three ways:

1. No difference in gas production between the unit containing the test compound and the control unit.

2. A lower gas output in the unit containing the test material than in the control.

3. A higher gas output in the unit containing the test material than in the control unit.

Condition 1 represents the case where the test compound present neither inhibits the methanogenic bacteria or was being degraded by them. Condition 2 on the other hand, represents the situation where the presence of the test material at that particular concentration inhibits the methanogenic output. In Figure 3.3 a nickel concentration of 50 mg/l provides an activity ratio of approximately 86 percent which corresponds to a 14 percent reduction in the gas output. Therefore, in relative terms, the nickel present in a system of unacclimated (organisms that have never been exposed to the test material before) methanogenic bacteria inhibits their metabolic activity by 14 percent when at a concentration of 50 mg/l. Condition 3 represents the situation where the test material is not inhibitory to the organisms and at the same time either stimulates or is degraded by them.

Even though the main purpose of toxicity tests is to determine toxic concentration levels, in some instances biodegradation can also be predicted. The batch toxicity test is the only toxicity test that can determine toxic levels and also whether or not a compound is being biodegraded by the methanogenic organisms. In batch toxicity tests the control unit and the test unit are fed equal amounts of acetic and/or propionic acid at the beginning of the experiment. Since no further substrate is added through the course of the experiment and if no inhibition occurs in the test unit, then equal total gas amounts within experimental error should be achieved in both units. If the gas produced in the test unit is greater than that in the control, this means that the test compound is being degraded since it is the only other available carbon source left in the system.

The following are 3 methods that have been developed and used to date:

1. Anaerobic Warburg Respirometer
2. Anaerobic Toxicity Assay
3. Batch Toxicity Assay

The methods will be briefly described in the following subsections.

Anaerobic Warburg Respirometer

A description of the Anaerobic Warburg Respirometer and how it works was provided in Chapter 2. The description given applied to the respirometer being used to determine whether or not the test material was biodegradable. As stated earlier, the two methods are basically the same, with the exception that in a toxicity test the material tested is no longer the sole carbon source for the bacteria. This means that cultures of methanogenic bacteria developed on acetic acid or any other of the volatile acid intermediates are used as the seed organisms and the same volatile acid is used as the carbon source in the feed solution.

Additional information on the anaerobic warburg respirometer is presented in studies by McCarty, et al. (1963) and Van Den Berg, et al. (1974). A study by Hovious, et al. (1973) utilized the anaerobic warburg method to initially screen the effects of several different petrochemical organic materials on the anaerobic digestion process. In this particular study the seed organisms used were taken from ordinary digested sludge, and the main carbon source was the carbohydrates and other materials found in the sludge solids.

Anaerobic Toxicity Assay

As a result of the disadvantages that occur when using the anaerobic warburg method (as discussed in Chapter 2) the anaerobic toxicity assay (ATA) was developed by Owen, et al. (1979). The ATA method utilizes the theory and procedures developed by Miller and Wolin (1974). The reactor unit is a stoppered 125 ml serum bottle and contains a combination of nutrients, test material, seed organisms and a carbon source for a total volume of 50 mls. In experiments reported in the literature [Owen, et al. 1979] acetate and propionate were the principal carbon sources. Even though the concentrations were not explicitly given, through calculation of the reported data, acetate was initially present in the system at a concentration of approximately 750 mg/l while propionate was at 250 mg/l. In another study by, Stuckey, et al. (1980), 0.10 ml of ethanol was added to the serum bottle volume of 50 ml. Liquid and gas samples were extracted by syringe for analysis.

Batch Toxicity Assay

The batch toxicity assay [Parkin, et al. 1980; Parkin and Kocher, 1980a,b; Yang, et al. 1980; Parkin and Speece, 1980] is a modified version of the anaerobic toxicity assay, to study toxicity to unacclimated methanogenic bacteria. Unlike the ATA method, which uses both propionate and acetate as the main carbon sources, the batch toxicity assay uses only acetate and at a much higher overall concentration (3,000 mg/l added as potassium acetate). Like the ATA however, the batch toxicity assay is carried out in a stoppered serum bottle (although at a larger volume of 150 ml) with the seed organisms, nutrients, carbon source and test material. The syringe method is also employed in the batch toxicity assay to measure the total amount of gas produced and to take liquid and gas samples.

Semi-Continuous Assays

The next category of tests to be discussed are the semi-continuous assays. Basically there are 2 major differences that distinguish the semi-continuous tests from the batch toxicity tests, they are:

1. A solids retention time is maintained in the system

2. Substrate levels are renewed daily on a batch basis

In batch toxicity tests, a small quantity of organisms is placed into a digestion unit to utilize a finite amount of substrate in a short time period. The semi-continuous assay, which closely resembles the operation of a conventional digestion process, involves a 24-hour cycle of feeding the unit, wasting solids, and measuring the daily gas production levels. Unlike the batch tests, each day an amount of fresh substrate is added to the unit to replace the substrate used by the organisms from the previous day. In order to provide the unit with the desired solids retention time, an appropriate amount of solids (active biomass) is removed daily from the system.

As was the case with batch toxicity tests, the seed organisms used in initially inoculating the units are developed in separate units from anaerobically digested sludge. These units are fed the substrate to be used in the study and after a period of several retention times (at which the units are maintained) the original population of organisms is replaced predominantly by a population of organisms that utilize the substrate being added as its main carbon source.

To start the experiment, the test units are inoculated with seed organisms, fed the proper nutrients and carbon source required for growth, and maintained at the proper temperature much in the same manner as the batch toxicity tests. Before the addition of the test compound, the test units should be operated for a period of at least 7 times the solids retention time and fed substrate amounts that will be used in the study in order to achieve a stable and reproducible operation (aproximately equal gas production among test units) [Stuckey, et al. 1980]. Following this procedure, the growth rate of the microorganisms is controlled by the solids retention time maintained in the unit and will eventually result in the development of a population of microorganisms that will be able to utilize the majority of the substrate added daily.

The compound to be tested is added to the units in amounts that will produce the desired concentrations at the same time as the fresh substrate feed solution is supplied. The toxicant in some instances may be added with the substrate or separately immediately after the substrate is added. In similar fashion to the batch toxicity tests, several reactors may be utilized to test the effect of the material over a range of concentrations.

There are two methods that can be used to determine the relative amount of inhibition caused by the presence of the test material. As is the case with most anaerobic toxicity tests, the effects of the test compound on the methanogenic bacteria present in the system is the principal objective of the test. The metabolic parameter used to monitor the methanogenic bacteria's activity is, once again, methane gas production.

The first method used to determine the relative amount of inhibition occurring in the system does not use a control as outlined in the batch toxicity test section. In this method a substrate used by the anaerobic microorganisms that produces methane gas in a well defined stoichiometric relationship is required. Solids are wasted on a daily basis in order to maintain the proper solids retention time, and fresh substrate is added in an amount only to replace that used from the previous day and to maintain a constant substrate concentration in the system. The steady-state uninhibited gas production (without any test compound present) is determined at the beginning of the experiment and is the control amount of gas production. The amount of added substrate that is needed can be determined by the amount of methane gas produced for that day, which is based upon the proper stoichiometric relationships. When a test material is added to the system and the gas production is less than the previously determined steady-state value, the system is considered to be inhibited.

The second method involves using the conventional control unit, when the added substrate is of a complex nature and the stoichiometry is not obvious. Sewage sludge is a good example of a complex substrate, since its exact composition is hard to ascertain and varies from treatment plant to treatment plant. Due to its variable composition, samples of the same volume from different plants often result in different methane gas production values. The only accurate procedure to measure the relative amount of inhibition when using complex substrates is to compare the performance of units containing test materials with that of a control unit when both are receiving equal daily amounts of substrate.

Semi-Continuous Serum Bottle Assay

The semi-continuous serum bottle assay developed by Parkin, et al. (1980) at Drexel University is an extension of the batch toxicity assay. A 24-hour cycle of feeding, wasting and measuring the daily gas production is included in the semi-continuous serum bottle method. The major differences between the 2 methods is that a solids retention time is maintained in the serum bottle by removing an appropriate amount of solids by means of a preset syringe, and substrate levels are returned to the experimental level (1050 or 2100 mg/l) to start a new 24-hour cycle.

To end a 24-hour cycle and start a new one, the amount of gas produced from the previous day is measured. Following the daily gas production measurement, glacial acetic acid (used as the test's substrate material) is added to replace that consumed from the previous day and bring its level back to the normal experimental value.

Consumption of acetic acid can be estimated from the following equation:

$$CH_3COOH \rightleftarrows CH_4 + CO_2 \qquad (3.4)$$

From the above stoichiometric relationship, equal amounts of methane gas (biologically) and carbon dioxide gas (through chemical neutralization) are produced. The amount of carbon dioxide produced is estimated from the above stoichiometry and from knowing how much acetic acid was added from the previous day. The methane production in turn is calculated by subtracting the CO_2 production from the total gas measurement for that day. The amount of acetic acid required to replace the previous day's loss is calculated from the methane production and equation 3.4.

Toxicants to be tested are added in slug doses (total toxicant added once per daily cycle) at the start of each new 24-hour cycle in order to minimize substrate utilization prior to toxicant exposure. Daily gas production was used as the monitoring parameter for measuring toxic levels.

In Figure 3.4, an example of a response to toxicant (nickel) exposure in a semi-continuous system is shown [Parkin, et al. 1980]. Interestingly, the systems which exhibited some inhibition (at 80-100 mg/l nickel) had about a 4 day lag period before the gas production dropped. However, all of the systems which were affected did recover within 10 days of the onset of the inhibition.

Semi-Continuous Control Method

The second method which relies on the use of a control unit to determine relative inhibition levels is represented by the procedure developed by Stuckey, et al. (1980). The method utilizes 1.5 liter continuously stirred conventional digesters operated at a 15 day solids retention time and a temperature of 35°C.

The digesters are initially seeded with anaerobically digested sludge. An equal mixture (on a COD basis) of primary and waste activated sludge or primary sludge alone was used as the sole substrate. In order to ensure that any effects observed in this test were due solely to the different concentrations of the test compound, a sufficient quantity of sludge (to be used as substrate) to last the entire test period should be collected, and stored at 2°C. As stated earlier, digesters are operated for a period of at least 7 days at the proper solids retention time and temperature so that a stable operation (similar gas production) can be achieved in all units. The solids retention time in this method was maintained at 15 days.

The test compound in an aqueous solution is added to the sludge just prior to its being fed into the test digester. With some test substances however the substances must be diluted in ethanol before 0.1 ml of this solution is added to the sludge before placing it in the digester. When the test compound is being diluted with ethanol, ethanol alone is added to the sludge being fed to the control units. As with the other, previously

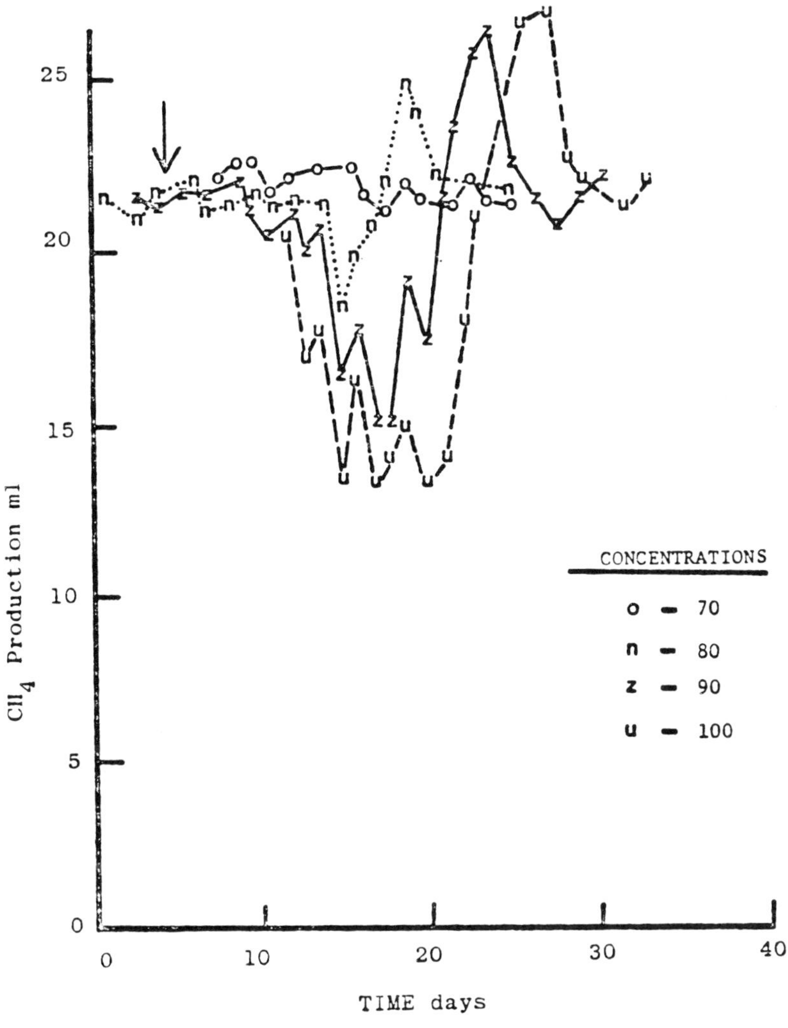

Figure 3.4 S.R.T. Response to Toxicant-Nickel, 50 Days at 35°C (Parkin, *et al*. 1980)

described, toxicity tests, the test compound is added to different units in varying amounts so that the effects of the test compound can be studied over a range of concentrations.

As an example, Figure 3.5 [Stuckey, *et al.* 1980] shows the effects that methylene chloride has on an anaerobic digester operating on a semi-continuous basis at a solids retention time of 15 days and a temperature of 35°C, in the concentration range of 0.625 to 10.0 mg/l. A concentration of 0.625 mg/l of methylene chloride had no adverse effect on the anaerobic organisms in the digester, since its gas production was equal to that of the control digester. A concentration of 2.5 mg/l of methylene chloride did inhibit the organisms slightly (as seen by the 20 percent decrease in gas production), but the system recovered to that of the control's gas production level after an 8 day period. The effects of 10 mg/l of methylene chloride injected into the anaerobic digester were much more severe. Gas production in the test unit was only 20 percent of that in the control unit. As was the case with the unit receiving 2.5 mg/l of methylene chloride, the unit containing 10 mg/l of methylene chloride recovered to produce gas amounts equal to the control but only after a 50 day period. It was believed that the anaerobic organisms were able to acclimate themselves to methylene chloride since, at the end of the study, the unit which was fed 10 mg/l of the material contained about 1 percent of the material added. It was also felt that although methylene chloride was toxic to the organisms, the material could be biodegraded by the organisms. The explanation for this phenomenon stems from the fact that the anaerobic organisms were able to acclimate to methylene chloride after a suitable time period and biodegrade the methylene chloride to concentrations below the toxic level [Stuckey, *et al.* 1980].

Reversibility Assays

Studies on reversing inhibition and toxicity effects in anaerobic fermentation systems were conducted by Parkin, *et al.* (1980). Experiments are initiated by injecting the toxicant at different concentrations into serum bottles. After the desired exposure time to the toxicant (for example, 1 hour, 1 day) the serum bottle was placed into a centrifuge to separate the active biomass from the

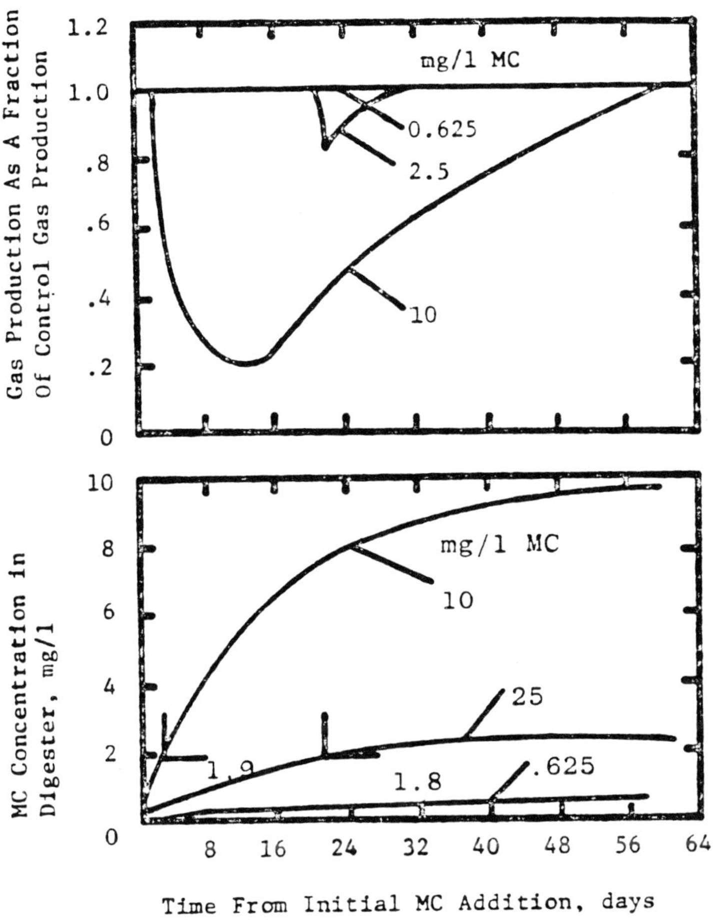

Fig.3.5 Upper:Normalized daily gas production versus time for methylene chloride feed. Calculated methylene chloride accumulation in digester versus time assuming no decomposition or loses (Stuckey, *et al*. 1980)

toxicant. Immediately following, toxicant free acetate enrichment media was injected into the serum bottle to replace the withdrawn supernatant. Daily feeding and gas measurement were then continued as outlined in the semi-continuous mode section for the purpose of monitoring the recovery of the system.

In the results of their studies, Parkin _et al_. (1980) found that a recovery from toxicant exposure was dependent on both toxicant concentration and exposure time. As shown in Figure 3.6 anaerobic systems exposed to nickel concentrations less than 800 mg/l and an exposure time of less than 1 day resulted in only a slight decrease in gas production once the toxicant in the liquid phase is removed. In addition, they found that a concentration of 2,400 mg/l (1 hour exposure) and an exposure to 800 mg/l for 4 days resulted in zero gas production for the duration of the experiment.

Continuous Assay

In addition to the above studies, Parkin _et al_. (1980) conducted continuous flow experiments using laboratory-sized anaerobic upflow filters containing void volumes of one liter. The filters were fed synthetic solution containing inorganic salt nutrients, buffers and acetate as the only carbon source. The loading to the filter was initially 133 lbs/ft^3/day (2,000 mg/l influent acetate concentration) which was later increased to 200 lbs/ft^3/day (3,300 mg/l influent acetate concentration).

For continuous addition toxicant studies, the toxicants were added at the desired concentrations in the daily feed. During 1 time slug addition studies, the feed solution contained the desired concentration for 1 day only. Afterwards, on the following days, the filter received only toxicant free feed solution where gas production and effluent COD values were determined so that the recovery progress of the reactor could be monitored. Additional slug doses (at higher concentrations) were then added and the above procedure repeated. As an example of their studies, systems exposed to nickel are used.

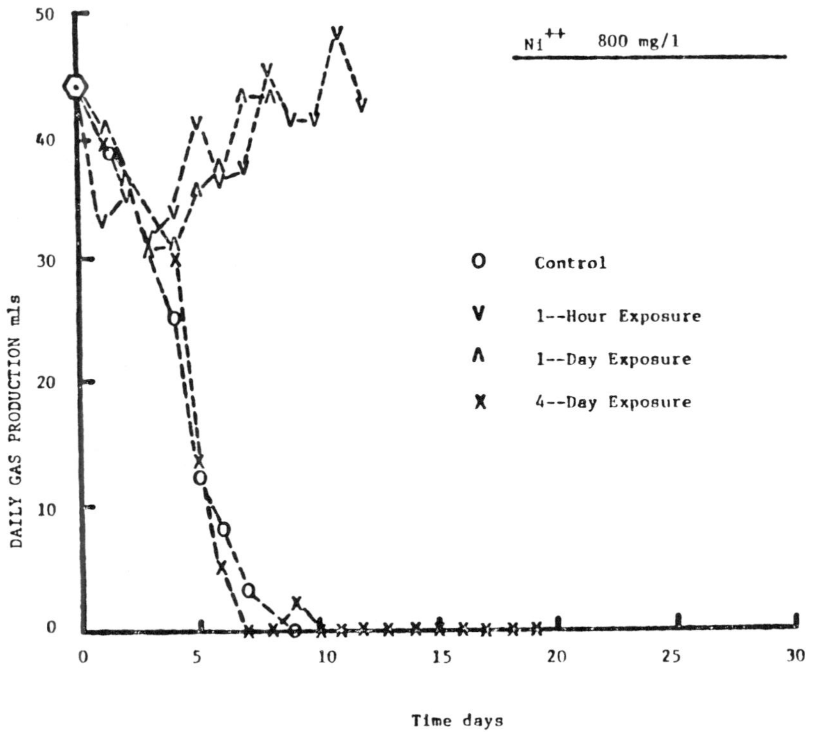

Fig. 3.6 Effect of Exposure on Toxicity Response:Nickel (Parkin, et al. 1980).

Figure 3.7 shows that continuous addition of nickel in the feed to an anaerobic filter for over 30 days indicated that once acclimated, 250 mg/l of nickel could be tolerated. One day slug additions of nickel were made to another anaerobic filter as depicted in Figure 3.8. At slug doses of up to 500 mg/l, only a 25 percent decrease in gas production was observed. After a 1,000 mg/l dose however, gas production almost entirely ceased, with the system not recovering significantly within five days.

Evaluation of Toxicity Tests

The toxicity tests described above were presented (with the exception of the reversibility test) in a sequence that was based on increasing solids retention time maintained in the test reactors of the various tests. This was done in order to illustrate the effect that the solids retention time and other environmental conditions have on the anaerobic microorganism′s ability to handle the toxic compounds being introduced into the system.

To accurately predict the response of the anaerobic microorganisms to the presence of a toxic substance, it is apparent that a toxicity test should, as closely as possible, simulate the conditions, including the solids retention time, experienced in full-scaled commercial processes. By doing so, factors that alter the degree of inhibition experienced in the full scale-scale processes such as acclimation, adaption, biodegradation and volatilization should affect the organisms in the test units to the same degrees. As an example, the anaerobic semi-continuous toxicity test′s mode of operation closely resembles that of the full-scale conventional digestion process and the extent and manner in which the above listed factors affect the organisms in the conventional process should occur in the semi-continuous test reactor.

The first set of tests presented were the batch toxicity tests: the anaerobic warburg respirometer, the anaerobic toxicity assay and the anaerobic batch assay. Unlike the semi-continuous and continuous toxicity tests, the value of batch toxicity tests is severely limited when used as a method for predicting system behavior when exposed

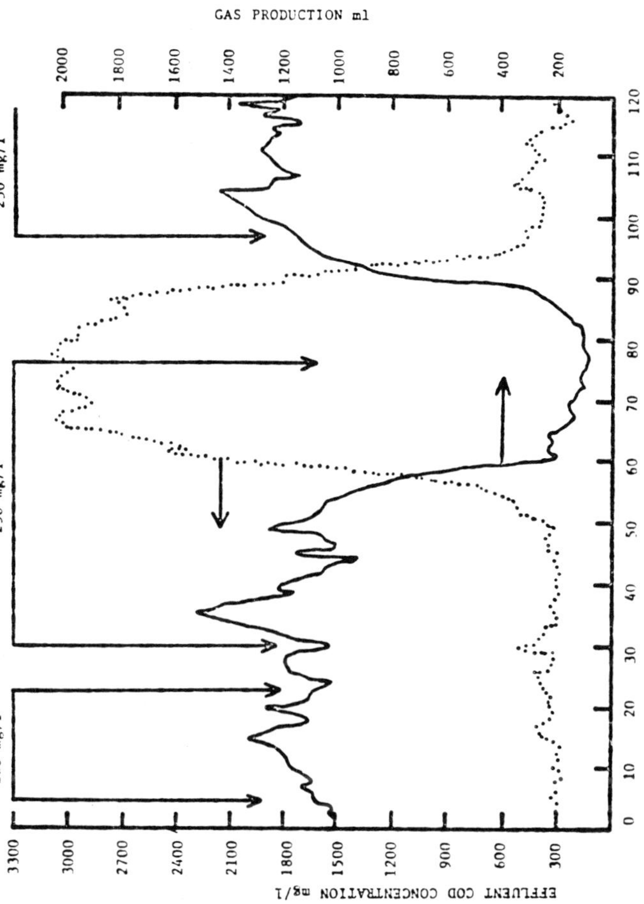

Fig. 3.7 Response of Anaerobic Filter to Toxicant Ni^{++} (Parkin, et al. 1980)

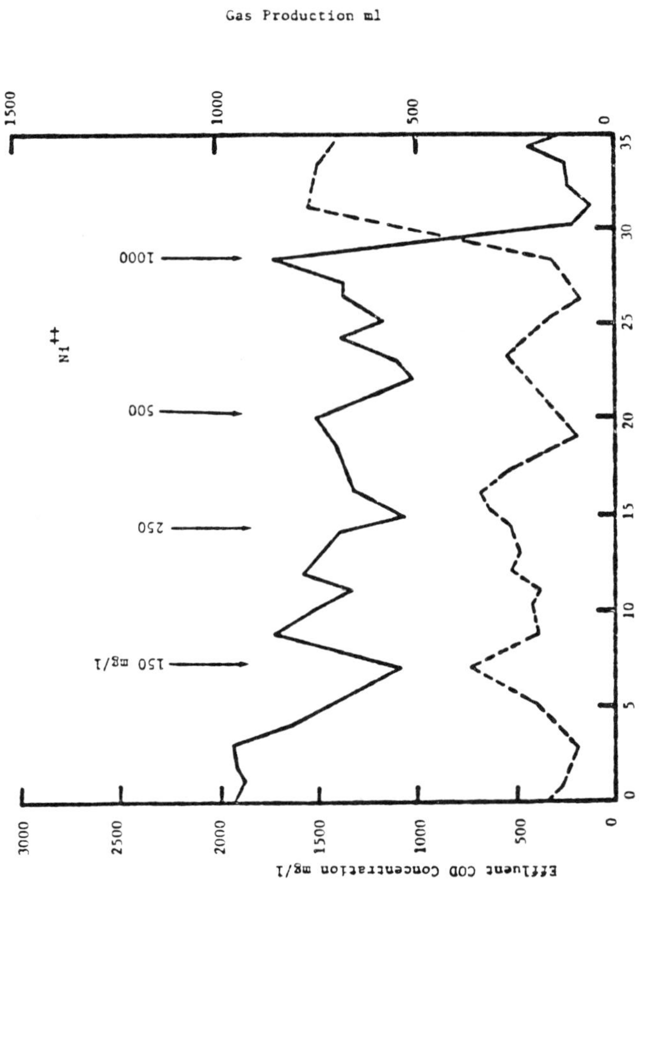

Fig. 3.8 Response of Anaerobic Filter to Toxicant (Parkin, et al. 1980)

to toxic materials and determining toxic concentration levels, since batch tests (their mode of operation) are not representative of any continuous flow process.

To illustrate this point, a study was conducted by Stuckey, et al. (1980) to compare and evaluate the differences in toxicity response to the conventional digestion process as predicted by the batch and semi-continuous toxicity tests. They found that the toxicity threshold (the concentration of the toxicant at which inhibition of the system starts to occur) was similar for both the batch and semi-continuous toxicity tests until acclimation, adaption, biodegradation and/or volatilization had occurred. If one of the above listed factors occurs in the system, then the semi-continuous digester is able to tolerate higher toxicant concentrations than the batch toxicity test had indicated. The main advantage of the semi-continuous toxicity test over the batch toxicity test is that it is able to take into account factors which predict the response of a full-scale conventional digestion system and toxic threshold concentration levels more accurately.

According to Stuckey, et al. (1980) the main advantages in using a batch toxicity test are that they are quick, relatively inexpensive and highly reproducible. Batch toxicity tests are able to provide information within 5 to 10 days after start-up. Also, long equilibrium periods are not required before start-up, as is the case with semi-continuous and continuous studies. The amount of labor required for a batch test is considerably less than with the other tests. Stuckey, et al. (1980) estimate that by using the anaerobic toxicity assay (or batch toxicity assay), 1 day is required to start 50 bottles plus an additional 7 to 10 hours are required for gas production monitoring over the 10 day study period. The semi-continuous method, on the other hand, using laboratory-sized digesters, requires 30 hours per week to feed and monitor 13 digesters. Due to the time requirements, using the batch toxicity assay or anaerobic toxicity assay, would give the researcher more flexibility since a larger number of samples and a large variation in concentration could be handled with little additional time or effort. Despite these advantages, the tests are still only able to provide a conservative value of the toxicity threshold concentration that will occur in the conventional digestion process due to a toxicant addition.

The next test presented was the semi-continuous toxicity test. The semi-continuous toxicity test was able to provide a more accurate prediction of system response to toxicant exposure than the batch toxicity tests, since it closely resembles the mode of operation of the full-scale conventional digestion process. The main reason that the semi-continuous test is able to accurately predict the response of the conventional digestion process is that the test is able to provide a similar mode of operation (fill and draw) while the solids retention times maintained are also similar. In full-scale operation the normal range of the solids retention time maintained is between 10 to 30 days. The lower end of this range represents the duration of the batch toxicity assay. Long solids retention times are favorable when a toxicant has been introduced into the system, since it provides the organisms with enough time for mechanisms such as acclimation and/or adaption to occur, thus reducing the relative amount of inhibition.

Essentially, when a system becomes inhibited, the organisms lose their metabolic activity and that includes the ability to reproduce. When a system becomes totally inhibited, all metabolic activity stops as shown through zero gas production. Studies at Drexel University [Parkin and Speece, 1980] have shown that periods of zero gas production are not necessarily indicative that the methanoganic bacteria are dead. Instead, if the toxicant dose is not too high, the organisms are able to go through a period of acclimation and/or adaption. After the proper period of acclimation, the bacteria return to their normal metabolic activity as demonstrated through gas production. If the acclimation period is longer than the solids retention time maintained in the reactor, then all the viable organisms will be washed from the system before the organisms can recover since the growth of new organisms does not occur. As an example, Figures 3.3 and 3.4 illustrated the effects that the solids retention times have on the acclimation process, In Figure 3.3 the batch system was continually inhibited at concentrations greater than 50 mg/l of nickel and not one of the units was able to recover during the 8 day test period. In Figure 3.4, however, concentrations of 30 to 100 mg/l of nickel did inhibit the semi-continuous test reactor, but all systems were able to recover within 10 to 12 days.

The main disadvantage with using a semi-continuous test is that it requires an equilibrium period before the toxic material can be introduced into the system. Also, long periods of operation are required to gather data; in some instances, up to 4 retention times. Semi-continuous tests also require more time to feed and monitor the digesters than the test reactors in the batch toxicity test. Semi-continuous tests are not as flexible as the batch tests, since additional effort is required to test numerous materials and variation in concentration for each.

The main advantage of the semi-continuous test is that it permits evaluation of reduced toxicity that might result from acclimation, biodegradation or volatilization.

In general, to predict the response of the anaerobic digestion process to toxicant exposure, the toxicity test should simulate the full-scale process as closely as possible. For this reason, to accurately predict the response of toxicant exposure to continuous digestion processes, the continuous toxicity test much in the same manner as the semi-continuous toxicity test, should utilize laboratory-sized test reactors that closely resemble the full-scale reactor. To date, the only documented continuous toxicity tests have been conducted at Drexel University [Parkin and Speece, 1980]. As with the semi-continuous test (and the semi-continuous full-scale reactor) the solids retention time also plays an important role in the recovery of continuous anaerobic processes when exposed to toxicant materials. It is generally conceded that these new continuous flow anaerobic digestion processes provide solids retention times that are several magnitudes higher than those maintained in the semi-continuous processes. Young and McCarty (1968) reported solids retention times in excess of 100 days while they were developing the anaerobic upflow filter process. Gorgan and Obayashi (1980) while studying temperature effects on the start-up of high SRT anaerobic processes estimated the solids retention time to be greater than

100 days in an upflow anaerobic sludge blanket reactor for loadings up to 13.5 kg COD/m^3/day and 25°C.

The effect of high solids retention times maintained in the continuous digestion process is similar to that experienced in the semi-continuous digesters but to a greater extent. Due to the greater solids retention times maintained, the continuous digestion processes have a greater capacity to recover from toxicant exposure. The longer the solids retention times maintained in the system, the longer the system has to acclimate to the toxicant. This means that some concentrations that produce irreversible inhibition in the semi-continuous process (due to washout of all viable organisms) may be sustainable if applied to the continuous digestion process since there are more organisms to be washed out before complete system failure. Figure 3.7 showed the response of an anaerobic filter to nickel at concentrations as high as 250 mg/l. Continuous addition of 250 mg/l of the metal initially inhibited the filter to one-seventh of its activity before toxicity addition. After approximately 45 days the filter recovered to full activity, as shown through gas production values.

The disadvantages of the continuous toxicity tests are similar in nature to those of the semi-continuous toxicity test listed previously. The advantage of the continuous toxicity test is also similar in nature to that listed for the semi-continuous test except that the organisms have a longer period of time for acclimation, adaption, etc., to occur, due to the higher solids retention times maintained in the continuous reactors. Long solids retention times and how they encourage system recovery from toxicant exposure, will be further discussed in the toxicity modeling section of this chapter.

The final toxicity test that was covered in this chapter was the reversibility toxicity test. Like the continuous toxicity test, the only documented research on the reversibility test has been at Drexel University [Parkin and Speece, 1980]. The main purpose of the test was to determine the maximum concentration level and exposure time of the added toxicant that would exhibit reversible toxic effects once the liquid supernatant was separated from the biomass and removed from the system. During the experiments conducted at Drexel University, most of the materials used in the reversibility test exhibited that the inhibition was reversible to some degree.

The results produced by a test will not have significance if the procedure used in the test cannot be carried out in a full-scale anaerobic digestion process. In order for the results produced by the reversibility test to have any significance, a procedure must be developed so that the toxicity effects produced in full-scale systems by toxicant addition can be reversed.

TOXICITY MODELS

In addition to the experimental work, several individual efforts have been made to develop models to predict toxic response. According to Parkin and Speece (1980), models that describe toxicity in methane fermentation systems should account for observed data and provide insight on process behavior Also, most traditional models assume steady-state conditions. Models describing toxicity must attempt to describe unsteady-state behavior.

Basically, toxicity models can be classified as either empirically-based or kinetically-based. Empirically-derived models are developed by observing the response curve (e.g. gas production versus time) of a system after it has been exposed to a toxic material in concentration that produce inhibition. The model is then developed by mathematically fitting a best-fit line through the response curve. Kinetically-based models, on the other hand, attempt to explain system response due to toxicant exposure by trying to model the effects that the toxicant has on the metabolic activity of the microorganisms involved in the process. The empirical models will be discussed first.

Empirically Based Models

The first empirically based model developed describes the recovery pattern from slug additions of toxicants to anaerobic fermentation systems. As seen in Figure 3.9 (in this case cyanide was the toxicant added), the shape of the observed recovery curve is similar in shape to the Streeter-Phelps

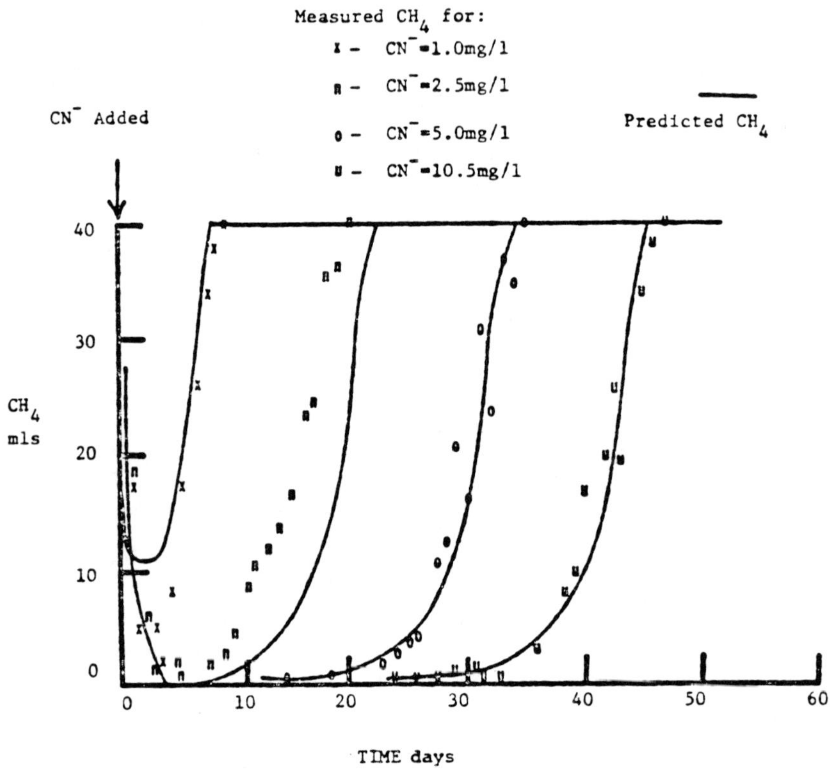

Fig. 3.9 Measured Versus Predicted Methane Production for Toxicant Exposure:Cyanide (Parkin and Speece, 1980)

Dissolved Oxygen Curve. According to Parkin and Speece (1980) the observed recovery pattern is remarkably similar for 30 industrial toxicants which they have studied. The list includes ions like ammonium, sodium, and sulfide, detergents, antibiotics, heavy metals and a wide variety of organics. It should be noted that this response curve is representative of a semi-continuous digestion system as developed by using the semi-continuous serum bottle technique described earlier.

In an effort to empirically fit an equation through the experimental data, an equation similar to the D.O. sag curve equation was chosen. The equation, developed by Parkin and Speece, (1980), is in the following form:

$$G_t = A_e^{-K_1 t} + B_e^{K_1 t} \qquad (3.5)$$

where:

G_t = methane production, ml/day

A,B = empirical constants

t = time after addition of toxicant, days

K_1 = "toxicity" rate constant, day^{-1}

K_2 = "acclimation" rate constant, day^{-1}

The constants A and B and the rate constants K_1 and K_2 are functions of toxicant type and concentration. Experimental data were used to develop the following relationships for the following constants:

$$K_1 = C(I_o)^d$$

$$K_2 = D(I_o)^d$$

$$B = E(I_o)^e$$

A = control methane production minus B

where:

I_o = concentration of toxicant in the serum bottle immediately following slug addition

C,D,E,c,d and e = constants determined from a regression analysis

The model has the ability to predict downtime (periods of zero gas production) as well as to predict the concentration that initiates inhibition (threshold concentration). In its present form, the model is strictly an empirical description of experimental data.

The second empirically based toxicity model was developed by Pearson, et al. (1980), in order to model the toxicity response of septic tanks when chemicals are introduced into the system that are toxic and inhibit biodegradation of organic wastes contained in the tank. The toxic chemicals are introduced into the septic tanks when recreational vehicles discharge their holding tanks into the septic tanks. These chemicals are present in the holding tanks to act as preservatives in order to control obnoxious odors.

Figure 3.10 is the response (as shown through gas production) of unacclimated septic tank sludge which was placed in the batch mode and dosed singly or in combination with toxicants used as preservatives (in this case formaldehyde, phenol and zinc) in the holding tanks of recreational vehicles. Any substrate or nutrients needed were considered to be already present in the sludge. The activity ratio, A, in the figure is defined as follows:

$$A = v/V \tag{3.6}$$

where: A = relative anaerobic gasification activity

v = total volume of gas produced from spiked sludge, ml

V = total volume of gas produced from unspiked sludge, ml

In Figure 3.10 the dotted line represents the actual activity ratios against increasing doses of toxicant. The solid S-shaped curves also shown in Figure 3.10 represent the best fit lines of the data presented. The equation for this line is as follows:

$$A = \left[1 + \sum_{i=1}^{m} \left(\frac{C_i}{K_i}\right)^a\right]^{-1} \tag{3.7}$$

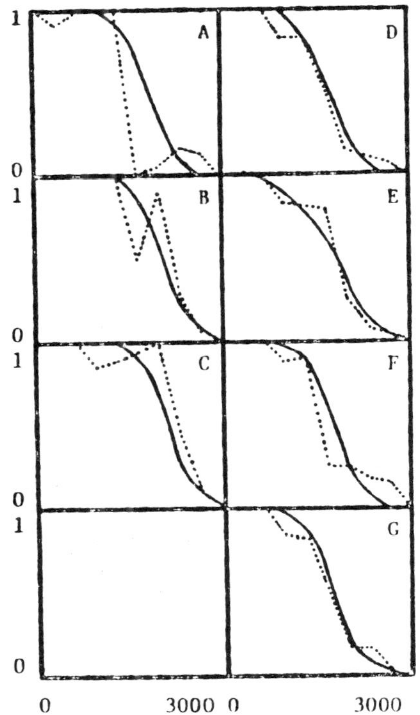

BLOCK	TOXICANT TYPE
A	Formaldehyde
B	Zinc Sulfate as Zinc
C	Phenol
D	Zinc and Formaldehyde
E	Phenol and Zinc
F	Formaldehyde and Phenol
G	Formaldehyde, Phenol and Zinc

Figure 3.10 Effect of Toxicant dose under Schock Loading on Anaerobic Gasification Activity (Pearson, et al. 1980)

where: m = number of preservatives

C_i = concentration of the ith preservative, mg/l

K_i = half kill dose, mg/l

= value of C_i for which A = 0.5 when m = 1

a = sensitivity exponent

The empirical equation that was arrived at using formaldehyde, zinc and phenol as toxicants is presented as follows:

$$A = \left[1 + \left(\frac{(\text{formaldehyde})}{200} + \frac{(\text{Zinc})}{400} + \frac{(\text{Phenol})}{500}\right)^{1.314}\right]^{-1} \quad (3.8)$$

The parenthesized chemical names represent concentrations, expressed in milligrams per liter. It should also be noted that equations 3.7 and 3.8 assume that the effects of combinations of toxicants to be strictly additive [Pearson, et al. 1980]. As noted previously, this equation was derived to model the response of unacclimated organisms operating in the batch mode of operation.

Similarly, attempts were made to model the effects of the chemical preservatives on septic tank sludge while operating in a semi-continuous reactor. In addition to different toxicant concentrations being tested, the semi-continuous reactors were also exposed to temperatures ranging from 9°C to 29°C. In order to test the effect of temperature on the anaerobic gas production rate versus temperature, plots were generated in the test for the different toxicants at concentrations of 30, 100, 300, 1000 and 3000 mg/l. Not surprisingly, their results showed that the gasification rate decreased with decreasing temperature (due to the lowering of the anaerobic organism's biological reaction rates) and increasing toxicant concentrations.

The anaerobic gas production rate in the semi-continuous test was computed as the hourly gas production per unit volume of sludge. The activity ratio, A, was also redefined to equal the rate of anaerobic gas production per unit volume of toxicant-spiked sludge, divided by the rate of anaerobic gas production per unit volume of unspiked sludge, at the same temperature.

The activity ratios are generated by equations 3.9 through 3.12 at any given temperature, and can be compared against the activity ratio of a control unit. Equations 3.9 to 3.12 are as follow:

$$A = \left[1 + \left(\frac{\text{(formaldehyde)}}{100}\right)^{0.2}\right]^{-1} \quad (3.9)$$

$$A = \left[1 + \left(\frac{\text{(zinc)}}{700}\right)^{0.5}\right]^{-1} \quad (3.10)$$

$$A = \left[1 + \left(\frac{\text{(phenol)}}{700}\right)^{0.5}\right]^{-1} \quad (3.11)$$

$$A = \left[1 + \left(\frac{\text{(quaternary ammonium) as Org. N}}{50}\right)^{0.25}\right]^{-1} \quad (3.12)$$

The chemical names in parentheses represent concentrations, in milligrams per liter.

Figure 3.11 is a series of curves that plot activity versus toxicant concentration for the batch and semi-continuous systems, as computed from equations 3.8 through 3.12. According to Pearson et al. (1980) the semi-continuous systems were found to respond more gradually to increasing toxicant dose than did the batch systems, which was reflected in a higher value of the sensitivity exponent, a, for the batch system.

Kinetically Based Toxicity Models

The first kinetically based toxicity model to be discussed was developed by Neufeld, et al. (1980) in order to determine whether or not anaerobic degradation was a viable process for the removal of phenol from industrial effluents.

At high substrate concentrations, the specific utilization rate, ds/xdt (as defined by equation 1.14) approaches a maximum value. It has been observed that for some substrates such as phenol and cyanide, increased concentrations lead to a maximum

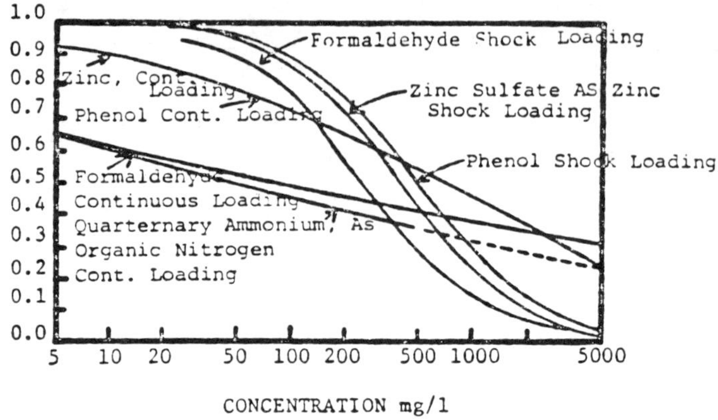

Figure 3.11 Effect of Toxicant Dose Under Shock and Acclimation Loading on Anaerobic Gasification Activity (Pearson, *et al*. 1980)

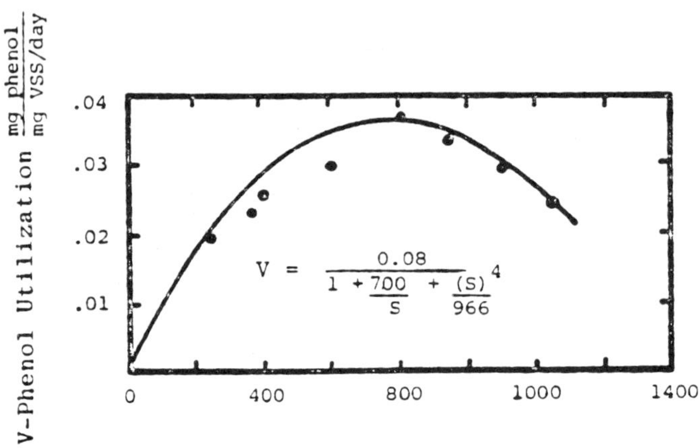

$$V = \frac{0.08}{1 + \frac{700}{S} + \left(\frac{S}{966}\right)^4}$$

Figure 3.12 Calculated relationship between specific rate of phenol utilization and residual phenol concentration overlayed with experimental data points. (Neufeld, *et al*. 1980)

in the specific growth rate (defined in equation equation 1.11) followed by a subsequent decline. This pattern was previously described in Figure 3.1. This type of relationship may be best described by a biokinetic model using the basic form of the Monod equation for substrate utilization, which was modified for substrate inhibition [Neufeld, *et al*. 1980]. The equation is as follows:

$$V = V_o / 1 + (K_s/S) + (S/K_1)^N \qquad (3.13)$$

where: N = order of inhibition

S = concentration of phenol, mg phenol/1

K_1 = inhibition constant, mg/1

V = rate of substrate utilization, mg phenol/mg VSS/day

V_o = theoretical maximum specific substrate utilization rate, mg phenol/mg vss/day

The procedure for determining V_o, K_s, K_1 and N will not be presented since it is explained in detail elsewhere [Neufeld, *et al*. 1980]. However, it should be noted that these constants were determined from the data obtained while running 3 10-liter semi-continuous laboratory test reactors. The ratio of hydraulic retention time to solids retention time was maintained at 3 to 1. The solids retention time varied from 30 to 120 days while the hydraulic retention time varied from 10 to 40 days. A synthetic solution was fed to the test reactors, with phenol as its only organic carbon source (and carbonate as the only inorganic source of carbon). The values of the constants as they were determined by the study were:

$N = 4$

$K_1 = 966$ mg/1

$K_2 = 700$ mg/1

$V_o = 0.08$ g phenol/gvss/day

The best-fit equation for the substrate inhibition model of anaerobic degradation of phenol was found to be:

$$V = 0.08/[1 + (700/S) + (S/966)^4] \qquad (3.14)$$

Figure 3.12 is a calculated plot of equation 3.14, where the solid dots represent actual data points. The calculated curve is identical in shape to the one presented in Figure 3.1, which was used to explain the different stages of toxicity. Point X on the curve presented in Figure 3.12 represents the point of maximum phenol utilization. By back calculating, the concentration of phenol at point X (where dv/ds= 0) was equal to 686 mg/l. Concentrations greater than 686 mg/l of phenol start to inhibit the microorganisms in the system.

A carbon balance on the system was taken before and at the conclusion of the study. It was assumed that phenol (organic) and carbonate (inorganic) were the only inputs of carbon into the system and the outputs from the system were new cell growth, acid production as volatile acids, CO_2 and methane gas. The carbon balance proposed by Neufeld, et al. (1980) is as follows:

a (phenol-C) + b(carbonate-C)= (3.15a)

c (biomass-C) + d(volatile acid-C) +

e (biological intermediate-C) + f(CO_2-C) +

g (methane-C)

At the conclusion of the experiment, the following carbon balance was obtained:

1.0 mg phenol-C + 0.315 mg carbonate-C (3.15b)

= 0.382 mg biomass-C + 0.261 mg volatic acid-C +

0.671 mg soluble TOC intermediate-C

From the balance, no methane gas was observed to be produced or any alkalinity utilized, indicating the reactions observed are for nonmethanogenic anaerobic organisms only and the model represents first stage anaerobic degradation of phenol [Neufeld, et al. 1980]. It was also found that a solids retention time of 40 days or greater is required for effective and stable degradation of phenol.

In an earlier article, Neufeld and Valiknac (1979) reported that aerobic phenol degradation in similar temperature ranges (32.2°C to 40.5°C) was equal to between 10.2 mg phenol/mg VSS/day and 14.7 mg phenol/mg VSS/day. The anaerobic value (0.08 mg, phenol/mg VSS/day) was only 0.05 - 0.8 percent of the aerobic specific rate of utilization.

The second kinetically based model to be discussed was developed at Drexel University [Parkin and Speece, 1980] and used the traditional Monod type equation as a starting point for developing a method of describing toxicity kinetics:

$$\frac{ds}{dt} = \frac{k\ S\ X_a}{K_s + S} \tag{3.16}$$

where: $\frac{ds}{dt}$ = rate of substrate utilization

k = specific utilization rate, mg acetate utilized per day per mg of methane bacteria.

X_a = concentration of active organisms, mg/l

K_s = Monod half velocity constant, mg/l acetate

As mentioned throughout the literature, conversion of acetic acid is considered the rate-limiting step in anaerobic digestion, and modeling acetate utilization should adequately describe the kinetics of methane fermentation of complex organics [Parkin and Speece, 1980]. Normal design procedure involves predicting minimum generation time, θ_c^{Min}. The two values are defined as follows:

$$\text{Minimum bacterial generation time} = \theta_c^{Lim} = (Yk-b)^{-1} \tag{3.17}$$

$$\text{Bacterial washout time} = \theta_c^{Min} = \left(\frac{Yk\ S_o}{K_s + S_o} - b\right)^{-1} \tag{3.18}$$

Where: Y = bacterial yield, mg bacteria synthesized per mg acetate utilized

b = bacterial decay rate, day^{-1}

The solids retention time used in design calculations of anaerobic systems calls for applying a safety factor to θ_c^{Min} or θ_c^{Lim}. Figure 3.13 shows the effects of θ_c, the solids retention time, and k, the specific utilization rate, on process efficiency by assuming steady state conditions for a continuously fed, complete mix reactor. The effects of K_s, the Monod half-velocity constant, is shown in Figure 3.14. Changes in Y and b also affect the process efficiency but not to the same extent as k and K_s.

Addition of toxicant materials changes the kinetics of methane fermentation. The first assumption of the model is that changes in the kinetic parameters originate from decreases in k and/or increases in K_s. According to Parkin and Speece (1980), the response to slug additions of toxicant (shown in Figure 3.15) can be modeled by approximating the unsteady state assumptions. By assuming that each day exhibits steady state behavior, algorithms for k and K_s can be developed as a function of time following (in slug dose) toxicant addition. Figure 3.16 shows the typical response of K_s and V_{max} (k x X_a) to the addition of toxic materials. As shown, V_{max} decreases while K_s increases as a result of toxicant addition. A return to non-toxic values of K_s and V_{max} can be obtained after the system goes through a suitable period of acclimation, metabolism, and toxic washout.

Once again, by assuming a steady state condition in a continuously fed, complete mix reactor, the combined effects of changes in k and K_s on process efficiency can be illustrated.

As an example, formaldehyde was used and is shown in Figure 3.17. At a formaldehyde concentration of 250 mg/l, the bacterial washout time temporarily increases from 7 days to 27 days. If the system had originally been designed for a solids retention time of 20 days, the systems could possibly fail because the washout time (due to toxicant exposure) exceeds the solids retention time. Whether or not a system fails depends upon the

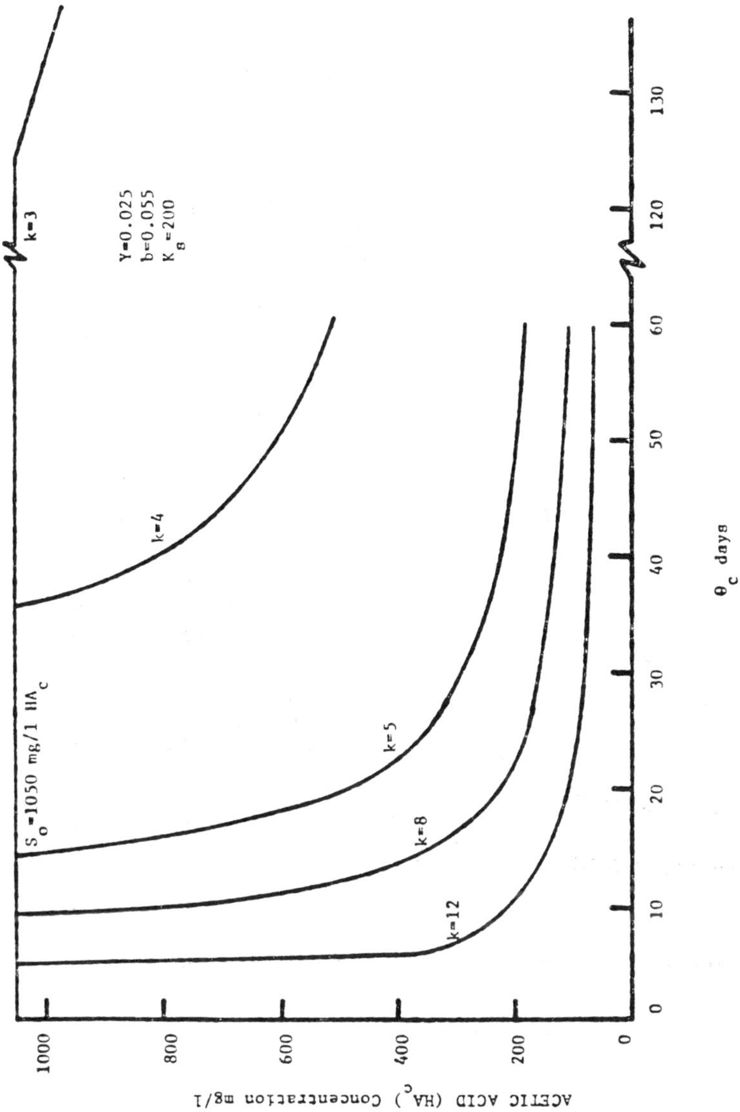

Figure 3.13 **Effect of k on Effluent HA_c Concentration (Parkin and Speece, 1980)**

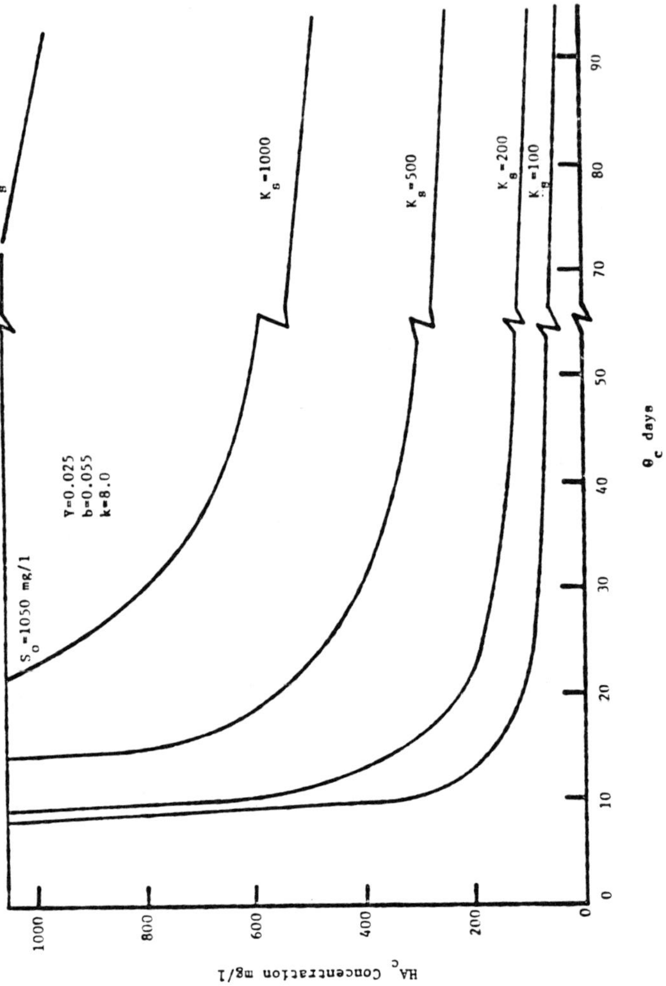

Figure 3.14 Effect of K_s on Effluent Acetic Acid Concentration (Parkin and Speece, 1980)

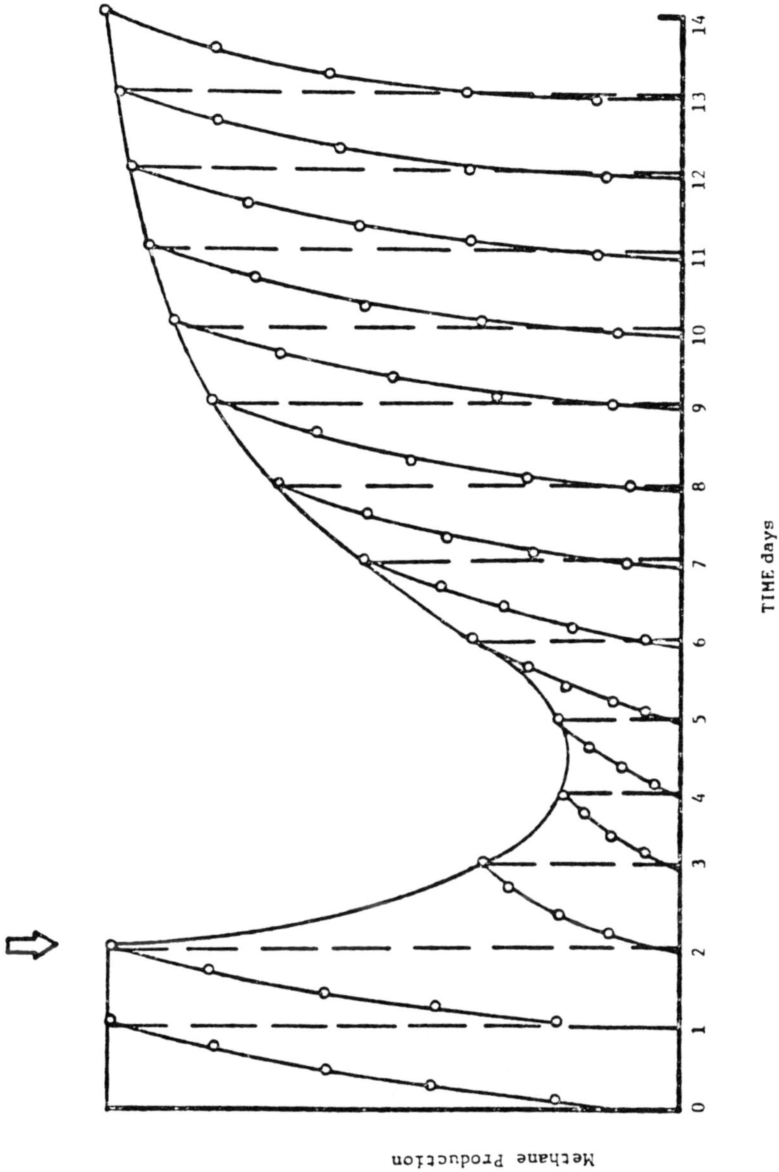

Figure 3.15 Daily Kinetic Evaluation Batch Fed, Continuously Operated Units (Parkin and Speece, 1980)

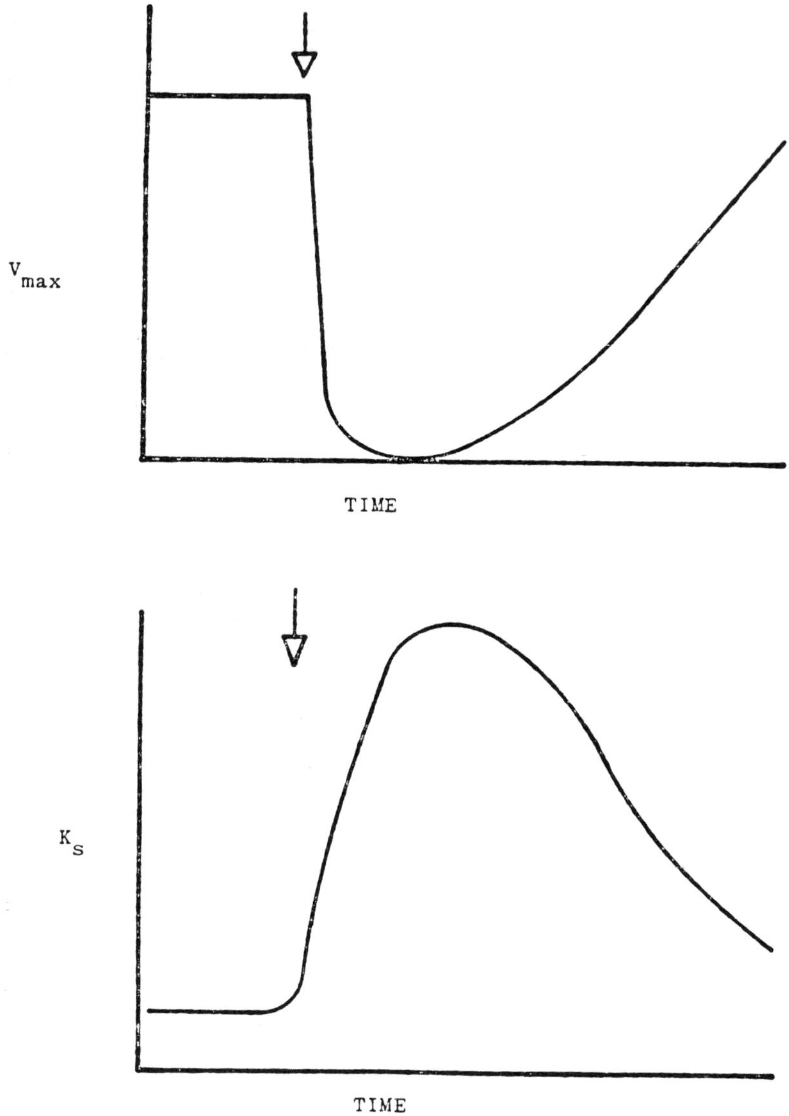

Figure 3.16 Acclimation Effect of Toxicant on V_{max} and K_s (Parkin and Speece, 1980)

toxicant type and concentration. Although Figure 3.17 was generated assuming steady state conditions, the approach can be used to describe unsteady state behavior. Figure 3.18 illustrates the effect of toxicant exposure on θ_c^{Min}. Once the toxic material is added, θ_c increases while process efficiency and methane production decrease. Even though Figure 3.18 was drawn as a steady state representation of θ_c^{Min}, it shows that θ_c^{Min} is seldom constant since it is subject to fluctuations dependent upon the stage of system recovery.

This kinetically based model points out the need for large SRT′s when treating industrial wastewaters containing chronic and transient toxic doses. Not only does a proper SRT value insure process stability, it allows for prolonged operation under conditions of washout prior to ultimate failure. The anaerobic filter, expanded bed, anaerobic rotating biological contactor and upflow anaerobic sludge blanket process all have the ability to provide high solids retention times and low hydraulic retention times. These reactors, therefore, have the built in capacity to withstand both transient and chronic toxicity better than other types of anaerobic processes.

The next model to be discussed, termed "Inhibition Coefficient Model," was also developed at Drexel University by Parkin and Speece (1980) to utilize the inhibition models developed by Lehninger (1976) and is based on enzyme kinetics. These equations combine inhibition coefficients with the substrate utilization equation. There are 3 basic types of reversible inhibitions described by Lehninger: competitive, non-competitive and uncompetitive. To this date, studies at Drexel University [Parkin, *et al*. 1980; Yang, *et al*. 1980; Parkin and Speece, 1980; Parkin and Kocher, 1980a,b] have shown toxicity from over 30 industrial toxicants to be reversible to some degree. The influence of inhibition on the substrate utilization equation is illustrated below [Parkin and Speece, 1980]:

Competitive: $$\frac{-ds}{dt} = \frac{kSX_a}{K_s(1 + T_x/K_I) + S} \qquad (3.19)$$

Non-Competitive: $$\frac{-ds}{dt} = \frac{kSX_a}{K_s + S(1 + T_x/K_I)} \qquad (3.20)$$

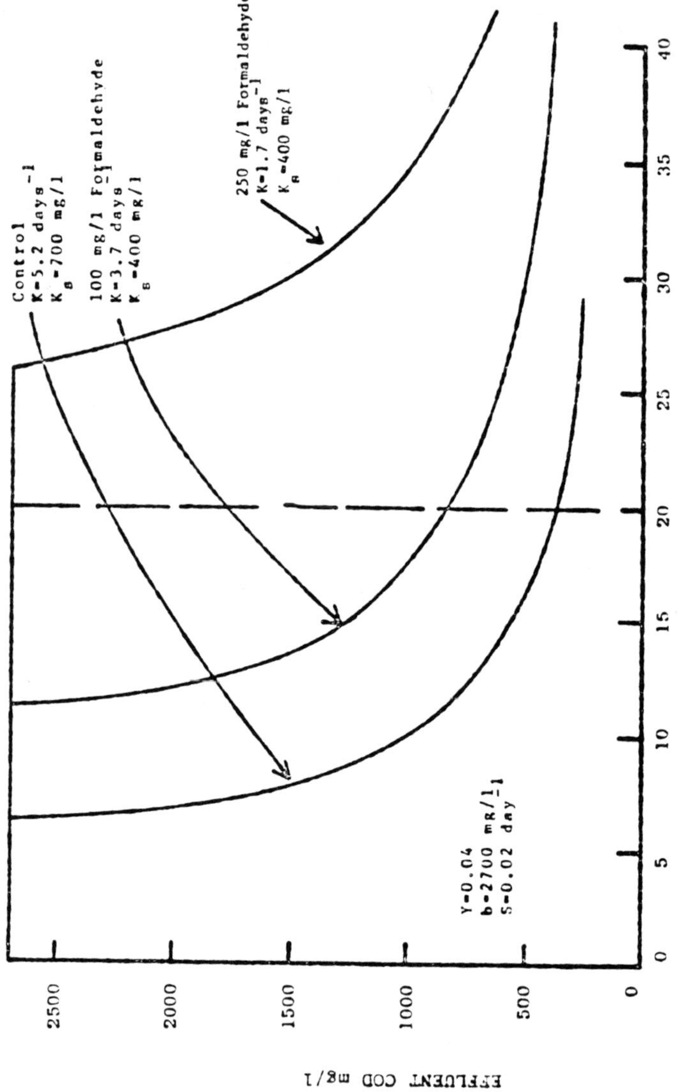

Figure 3.17 Steady-State Characteristics for Toxicity Kinetic Parameters-Formaldehyde (Parkin and Speece, 1980)

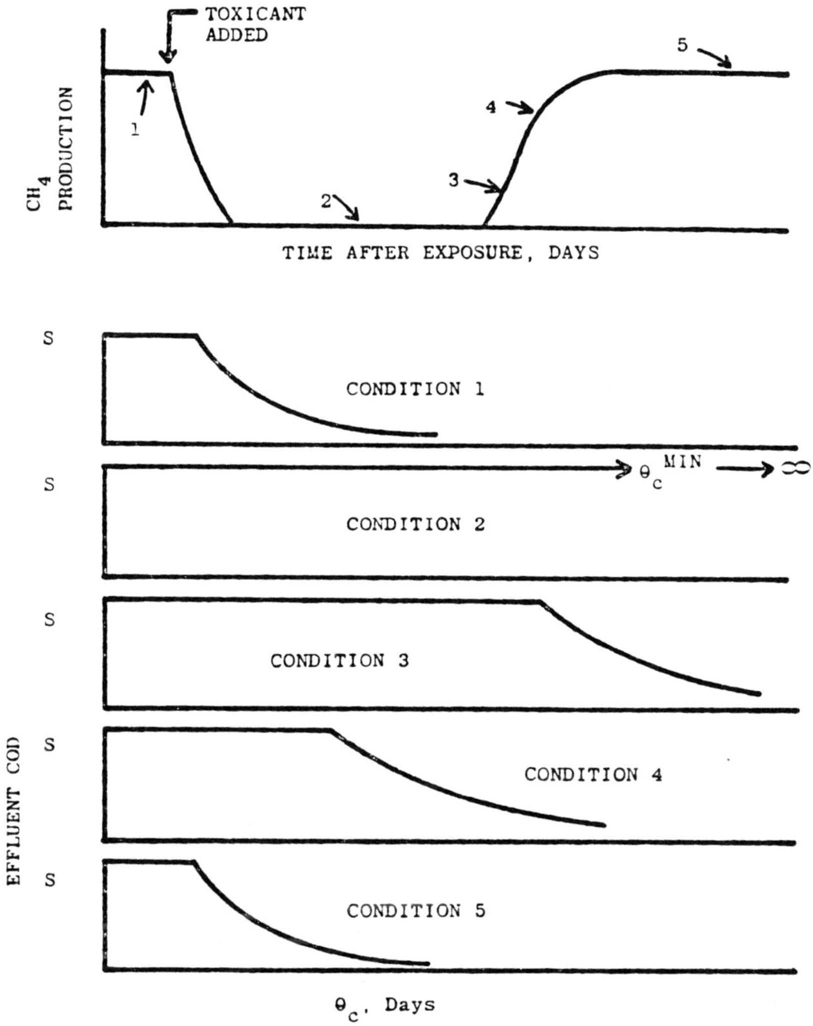

Figure 3.18 Effect of Transient Nature of Toxicity Response on Process Efficiency (Parkin and Speece, 1980)

$$\text{Uncompetitive:} \ \frac{-ds}{dt} = \frac{kSX_a}{(K_s + S)\ (1 + T_x/K_I)} \tag{3.21}$$

where:

T_x = concentration of toxicant, mg/l

K_I = inhibition coefficient (function of toxicant type), mg/l

From the above equations, changes in k are associated with noncompetitive inhibition while K_s changes are tied to competitive inhibition.

For slug doses of toxicants, the typical unsteady state reponse curve shown in Figure 3.18 can be generated in a similar manner used in the first model. After the addition of the toxicant, T_x and K_I values change due to acclimation, metabolism and washout. K_I values can be determined from experimental data by integrating the inhibition equations and solving for K_I. According to Speece and Parkin (1980), the shape of the K_I versus time curve should be identical to the gas production versus time response curve shown earlier. Also, the effect of chronic toxicity on system performance can be seen by assuming a steady state, continuously fed, complete mixed system and by varying T_x and/or K_I. The resulting equations are:

$$\text{Competitive:} \ S = \frac{K_s(1 + b\theta_c)\ (1 + T_x/K_I)}{\theta_c(YK - b) - 1} \tag{3.22}$$

$$\text{Uncompetitive:} \ S = \frac{K_s(1 + b\theta_c)}{\theta_c YK - b\ (1+T_x/K_I)(1 + T_x/K_I)} \tag{3.23}$$

$$\text{Non-Competitive:} \ S = \frac{K_s(1+b\theta_c)\ (1 + T_x/K_I)}{\theta_c YK - b(1 + T_x/K_I)(1 + T_x/K_I)} \tag{3.24}$$

Figures 3.19 and 3.20 show the effect of T_x and K_I on process performance for non-competitive inhibition. Similar procedures can be conducted for competitive and uncompetitive inhibition. K_I values can be determined from laboratory experiments using chemostats.

As in the previous model, an increase in θ_c^{Min} is caused by toxicant addition. If process efficiency is not to be severely decreased after

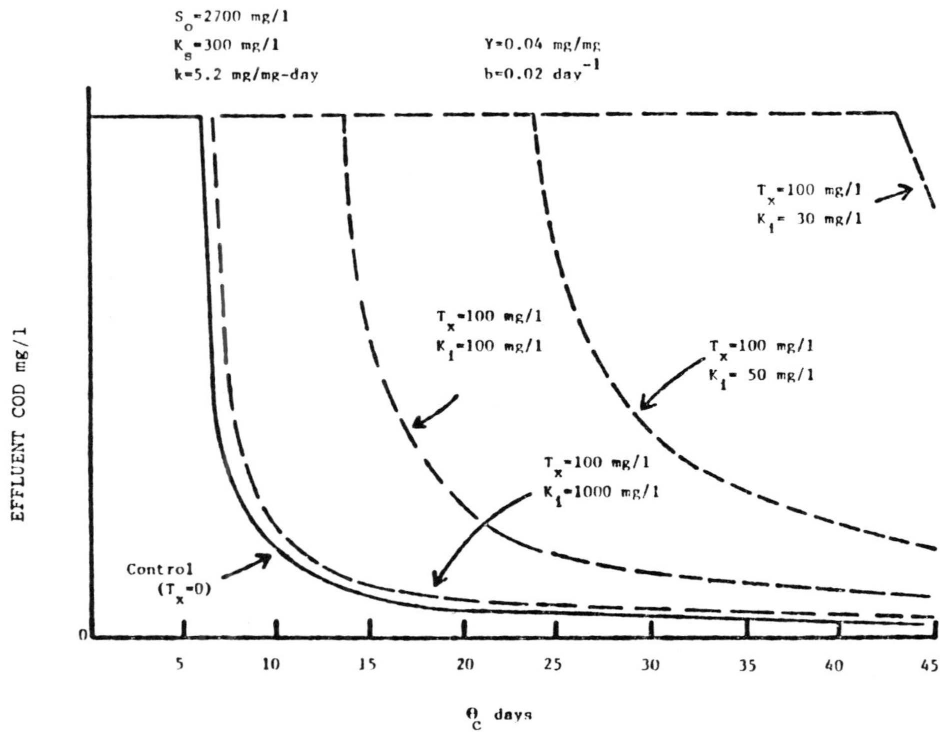

Figure 3.19 Non-Competitive Inhibition Effect of T_x on Effluent COD (Parkin and Speece, 1980)

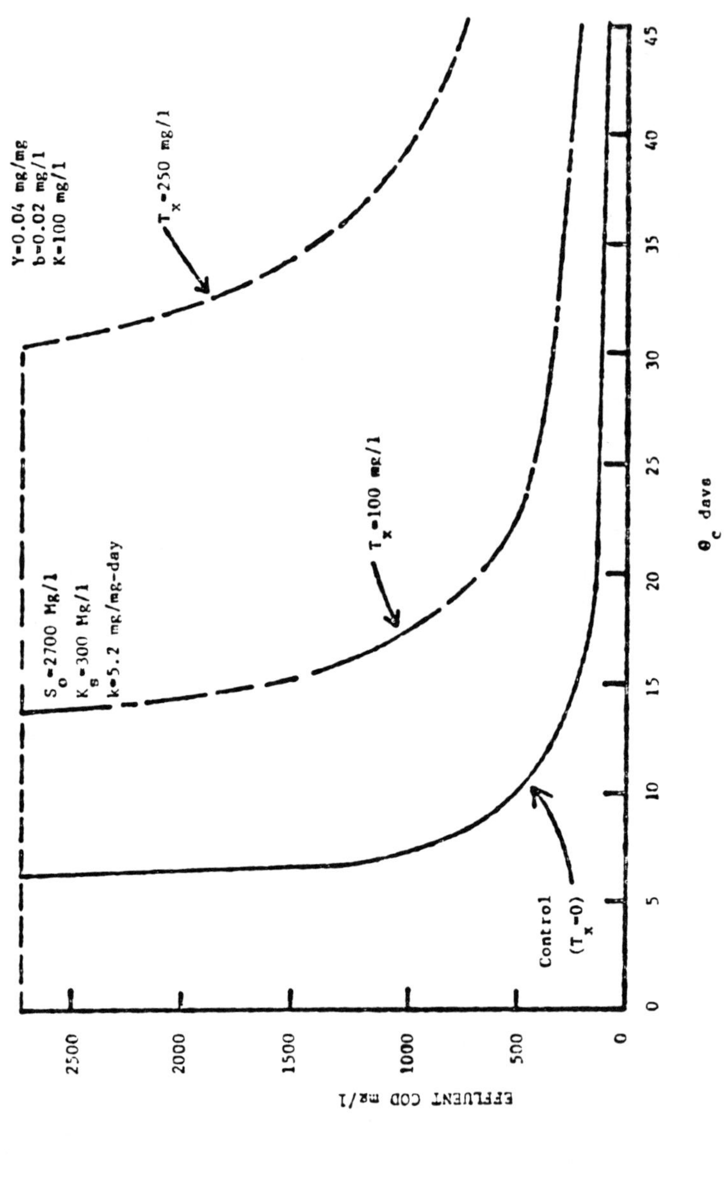

Figure 3.20 Non-Competitive Effect of T_x on Effluent COD (Parkin and Speece, 1980)

toxicant exposure, sufficiently long SRT values are required.

Evaluation of Models Predicting Toxicity

During the toxicity studies conducted by Parkin and Speece (1980); Parkin, et al. (1980); Yang, et al. (1980) and Parkin and Kochner (1980 a,b), 3 models that predict the response of an anaerobic system to toxicant addition were developed; 1 empirical model and 2 kinetically based models. From looking at the recovery patterns of over 30 common industrial toxicants, some of which were presented earlier, extended periods of little or no gas production resulting from toxicant exposure are not necessarily indicative of destruction of methanogenic viability. Gas production representing microbial viability returns to normal through acclimation, metabolism and toxicant washout.

From a design and operational point of view, the results from the Drexel University recovery models as well as the experimental data indicate the importance of the solids retention time maintained in the system. In order to obtain the acclimation results displayed in the studies, sufficiently large design SRT's are required. Large SRT values allow recovery from temporary slug doses of toxicants by preventing total washout of biomass prior to recovery which takes place through acclimation and/or washout of the toxicant to levels that can be tolerated by the system. The relatively new processes such as the upflow anaerobic filter, expanded fluid bed, rotating biological contactor and the upflow anaerobic sludge blanket process provide high SRT values at low hydraulic retention times and inherently offer the most promise for effectively treating industrial wastewaters containing toxicants.

Two models presented, unlike the models developed at Drexel University, were developed by observing the response of an anaerobic system from the addition of only a few different toxicants. The model developed by Pearson, et al. (1980) empirically fit the response of septic tank sludge (in batch and semi-continuous systems) with a best fit line from the addition of toxicants used as chemical preservatives for odor control. Even though only 4 toxicants were studied (zinc, phenol, formaldehyde and ammonia), the procedure used could be readily applied to other toxicants.

One significant drawback of the Pearson study was that little emphasis was placed on the effect that the solids retention time maintained in the test systems had on the response of the system. Basically, the model empirically generated a best fit line for the activity versus concentration plot derived from batch and semi-continuous systems data. One possible explanation for ignoring SRT effects was that the goal of the study was to determine and recommend maximum concentrations of preservatives that could be tolerated in the holding tank without severely inhibiting the organisms inside septic tanks (since the holding tanks are emptied into septic tanks). The study did, however, include several semi-continuous unit runs, which allowed the organisms the chance to acclimate to the toxicants. With the data generated from the semi-continuous systems, the model constants were recalculated to include acclimation, allowing the values generated from each system to be compared.

The purpose of the study by Neufeld, et al (1980) was to determine if anaerobic degradation was a viable process for the removal of phenol from industrial wastewaters. As a result, a best fit curve identical in shape to Figure 3.1 which describes the relative stages of inhibition for any material, was used and calibrated by experimental data. The experimental data was obtained by running semi-continuous reactors at several hydraulic and solids retention times. As shown in Figure 3.12, the calibrated best fit model curve closely fits the experimental data.

A significant finding of that study was that solids retention times greater than 40 days are required for stable anaerobic phenol utilization. It was further suggested, as in the case with the Drexel University studies, that systems which provide high solids retention times such as the anaerobic filter or anaerobic contact process would be suitable for phenol degradation. In the model, it was determined that no methane gas was produced. The application of this method would be a pretreatment step, since phenol is reduced to intermediate levels that are readily degradable by anaerobic treatment.

It should not be assumed that the curve generated in Figure 3.12 is unique to the anaerobic degradation of phenol. As stated previously, there is a wide range of materials from cations to organic materials that, when introduced into an anaerobic system in increasing concentrations, would result in the same general curve. The procedure described by Neufeld, et al. (1980) can be used to generate similar results with other materials.

LISTING OF COMPOUNDS KNOWN TO BE TOXIC TO ANAEROBIC SYSTEMS

The final section of this chapter summarizes the available concentration values of various compounds that have been documented (through the literature) to severely inhibit the various anaerobic waste treatment processes. Table 3.3 provides a summary of the available toxicity concentrations for 37 common metallic and organic compounds and is arranged according to the hydraulic flow scheme of the various anaerobic processes. Table 3.3 was developed not only to provide toxic concentration levels for the different processes but also to show the effects of testing under different environmental conditions (i.e., hydraulic flow scheme, substrate, substrate concentration, temperature, etc.).

In the first column of Table 3.3, the names of the compounds that have been documented in literature to inhibit the anaerobic digestion process are listed in alphabetical order, while the second column provides the toxicant concentration that inhibits the metabolic activity (when compared against the gas production from a control unit) of the organism by 50 percent in a batch reactor. The concentration values were determined by the batch toxicity tests, e.g., the anaerobic respirometer, batch toxicity assay and the anaerobic toxicity assay, all of which were described previously. In addition to providing the toxic concentration for the various compounds, the reference from which the value was taken is provided directly beneath the concentration value. This procedure is used for all concentration values listed in Table 3.3.

The third column of Table 3.3 lists toxicant concentrations that inhibit the conventional digestion process. The values provided are those concentrations that produce total process failure,

Table 3.3 Comparison of Concentration Levels that Cause Toxicity to the Different Anaerobic Treatment Processes.

Toxicant	Toxic Concentration Levels That Causes 50% Inhibition in Batch Systems (mg/l)	Toxic Concentration Levels[1] for Semi-Continuous Systems (mg/l)	Toxic Concentration Levels[2] for High SRT Semi-Continuous Systems (mg/l)	Toxic Concentration Levels[2] for Continuous Systems (mg/l)
Acrolein	11.5 (Lin Chou, et al. 1978)	-	-	-
Acrylonitrile	100 (Hovious, et al. 1973)	5[3] (Barth and Bunch, 1978)	-	-
Ammonium	7,500-10,000 (Parkin, et al. 1980)	-	12,000 (Parkin and Kocher, 1980)	6,000 (Parkin, et al. 1980)
Cadmium	-	1 (Barth and Bunch, 1979)	200 (Parkin and Kocher, 1980)	-
Calcium	8,000[3] (Kugelman and McCarty, 1964)	-	15,000-20,000 (Parkin and Kocher, 1980)	-
Carbon Tetrachloride	-	10[3] (Barth and Bunch, 1979)	-	-
Chloroform	-	0.1 (Barth and Bunch, 1979)	7.5 (Parkin and Kocher, 1980)	40 (Yang, et al. 1980)
Chromium III	-	-	100 (Parkin and Kocher, 1980)	-
Chromium IV	-	-	40-60 (Parkin and Kocher, 1980)	-
Copper	-	10 (Barth and Bunch, 1979)	60 (Parkin and Kocher, 1980)	-
Crotonaldehyde	50-100 (Hovious, et al. 1973)	-	-	600 (Hovious, et al. 1973)
Cyanide	1 (Yang, et al. 1980)	-	5 (Parkin and Kocher, 1980)	20 (Yang, et al. 1980)
1,2-Dichlorethene	-	1[3] (Barth and Bunch, 1979)	-	-
1,1-Dichloroethylene	-	-	50-100 (Parkin and Kocher, 1980)	-
Ethyl Benzene	340 (Lin Chou, et al. 1980)	-	1000 (Parkin and Kocher, 1980)	-
Ethylene Dichloride	50 (Stuckey, et al. 1980)	2-8[4] (Stuckey, et al. 1980)	-	-
Formaldehyde	200 (Pearson, et al. 1980)	100[4] (Pearson, et al. 1980)	150-200 (Parkin and Kocher, 1980)	400 (Parkin, et al. 1980)
Gasoline	-	-	5,000-7,500 (Parkin and Kocher, 1980)	-
Hyamine 1622	-	-	50-100 (Parkin and Kocher, 1980)	-
Hyamine 3500	-	-	50 (Parkin and Kocher, 1980)	-
Hydrazine	-	-	150 (Parkin and Kocher, 1980)	-

Table 3.3 (Continued)

Toxicant	Toxic Concentration Levels That Causes 50% Inhibition in Batch Systems (mg/l)	Toxic Concentration Levels[1] for Semi-Continuous Systems (mg/l)	Toxic Concentration Levels[2] for High SRT Semi-Continuous Systems (mg/l)	Toxic Concentration Levels for Continuous Systems (mg/l)
Lead	-	50 (Barth and Bunch, 1979)	-	-
Magnesium	50-70 (Kugelman and McCarty, 1964)	-	-	-
Methylene Chloride	14 (Stuckey, et al. 1980)	>10 (Stuckey, et al. 1980)	-	-
Nickel	300 (Parkin, et al. 1980)	40 (Barth and Bunch, 1979)	>100 (Parkin and Kocher, 1980)	>250 (Parkin, et al. 1980)
Nitrobenzene	12.5 (Lin Chou, et al. 1978)	-	-	-
Phenol	500 (Pearson, et al. 1980)	700[4] (Pearson, et al. 1980)	-	-
Potassium	12,000 (Kugelman and McCarty, 1964)	-	-	-
Sodium	8,000 (Kugelman and McCarty, 1964)	-	-	-
Sulfide	>400 (Parkin, et al. 1980)	200 (Lawrence, et al. 1964)	>600 (Parkin and Kocher, 1980)	>800 (Parkin and Kocher, 1980)
Tetrachloroethylene	80-160 (Benson and Hunter, 1976)	-	-	-
1,1,1-Trichloroethane	(Benson and Hunter, 1976)	1[3] (Barth and Bunch, 1979)	-	-
1,1,2-Trichloro-1,2,2-Triflouroethane	10-20 (Benson and Hunter, 1976)	-	50-250 (Parkin and Kocher, 1980)	-
Trichlorofluoromethene	-	0.7 (Barth and Bunch, 1979)	-	-
Vinyl Acetate	11.5 (Stuckey, et al. 1980)	>1,200[4] (Stuckey, et al. 1980)	-	-
Vinyl Chloride	40 (Stuckey, et al. 1980)	65[4] (Stuckey, et al. 1980)	-	-
Zinc	-	10 (Barth and Bunch, 1979)	-	-

1 Toxic concentration refers to concentration that causes process failure unless otherwise noted.
2 Toxic concentration is defined as no methane gas production for a period greater than 50 days.
3 Concentration that is contained in sludge being digested.
4 Represents concentration that produces 50% inhibition level.
> Represents a concentration higher than stated.
< Represents a concentration less than stated.

unless otherwise noted. The values listed were either measured experimentally or observed to cause complete failure in full-scale systems. As stated previously, for mesophilic digestion the normal solids retention time is usually in the range of 10-30 days. All values listed in the third column were obtained from systems operating in this range.

The major contributors to the values found in Column 3 were Barth and Bunch (1979). The concentration values listed under Barth and Bunch represent a combination of data which they collected from the literature, along with data determined from their own work. In their studies, Barth and Bunch utilized a 1.5 liter conventional digester (30 day SRT) using primary sewage sludge as the main substrate, to determine the effects of copper, nickel and zinc. All other values listed under Barth and Bunch represent the values they collected from the available literature.

The fourth column of the table provides toxicant concentrations that were determined to inhibit a semi-continuous anaerobic system maintained at a 50 day SRT by the semi-continuous serum bottle test developed by Parkin and Speece (1980). It should be noted that the term "Toxic Concentration," as defined by Parkin and Speece, is the concentration of toxicant that produced zero gas production in an anaerobic system for a period greater than 50 days.

The final column of Table 3.3 lists toxicant concentrations that produced toxic inhibition in a laboratory sized upflow anaerobic filter reactor. The toxicant was introduced into the reactor along with the influent feed solution on a continuous basis. The upflow anaerobic filter process is one of several new anaerobic treatment processes that are able to handle concentrated industrial effluents continuously and directly, since they are able to maintain high solids retention times independent of hydraulic flow. As was the case with the semi-continuous test, zero gas production for a period greater than 50 days was taken as indicative of process failure.

Table 3.3 was arranged so that reading the columns from left to right for any compound, the various anaerobic treatment systems are listed in terms of increasing solids retention times. This allows for a general comparison between the concentrations that cause toxic inhibition and the solids retention times maintained in the various anaerobic treatment systems, for the same compound. As demonstrated by Table 3.3, the general trend confirms the fact that systems which maintain higher solids retention times are able to handle a higher toxicant concentration before system failure. Formaldehyde, as shown in Table 3.3, is an example of this trend.

As can be seen in Table 3.3 numerous studies have contributed concentration values. Many of these studies were described in detail previously in this chapter.

REFERENCES

Barth, E.F. and Bunch, R.L., 1979. Biodegradation and Treatability of Specific Pollutants. Environmental Protection Series. EPA 600/9-79-034. U.S. Government Printing Office, Washington, D.C.

Benson, E.N. and Hunter, J.V., 1976. "Comparative Effects of Halogenated Hydrocarbon Solvents on Waste Disposal Processes." Paper presented at the 31st Purdue Industrial Waste Conference, Purdue University, West Lafayette, Indiana, May 1976.

Dickerson, R.E., Gray, H.B. and Haight, G.P. 1974. Chemical Principals. W.A. Benjamin, Inc., Menlo Park, California.

Gorgan, J. and Obayashi, A., 1980. Unpublished Data, Illinois Institute of Technology, Chicago, Illinois.

Hovious, J.C., Waggy, G.T. and Conway, R.A. 1973. Identification and Control of Petrochemical Pollutants Inhibitory to Anaerobic Processes. Environmental Protection Series. EPA R2-73-194.

Jeris, J.S. and McCarty, P.L. 1965. "The Biochemistry of Methane Fermentation Using C14 Tracers." J. Wat. Poll. Con. Fed. 37:178-186.

Keith, L.H. and Telliard, W.A. 1979. "Priority Pollutants: A Perspective View." Engr. Sci. Tech. 13:416-423.

Krisch, E.J., Grady, C.P.L. Wukasch, R.F. and Tabak, H. 1981. "Protocol Development for the Prediction of the Fate of Organic Priority Pollutants in Biological Wastewater Treatment Systems." Purdue University, West Lafayette, Indiana.

Kugelman, I.J. and Chin, K.K. 1971. "Toxicity, Synergism, and Antagonism in Anaerobic Waste Treatment Processes." In Anaerobic Biological Treatment Processes. Ed. R.F. Gould, pp. 55-90. Advances in Chemistry Series No. 105. American Chemical Society, Washington, D.C.

Kugelman, I.J. and McCarty, P.L. 1965. "Cation Toxicity and Stimulation in Anaerobic Waste Treatment." J. Wat. Poll. Con. Fed. 37:97-116.

Kugelman, I.J. and McCarty, P.L. 1965. "Cation Toxicity and Stimulation in Anaerobic Waste Treatment: II. Daily Feed Studies." Paper presented at the 19th Purdue Industrial Waste Conference, May, 1965, Purdue University, West Lafayette, Indiana.

Lawrence, A.W., McCarty, P.L. and Guerin, F.J. 1964. "The Effects of Sulfide on Anaerobic Treatment." Paper presented at the 19th Purdue Industrial Waste Conference, Purdue University, West LaFayette, Indiana, May 1964.

Lawrence, A.W. and McCarty, P.L. 1965. "The Role of Sulfide in Preventing Heavy Metal Toxicity in Anaerobic Treatment." J. Wat. Poll. Con. Fed. 37:392-406.

Lehninger, A.L. 1976. Biochemistry. pp. 197-206, Worth Publishers, New York.

Lin Chou, W., Speece, R.E., Siddigi, R.H. and McKeon, K. 1978. "The Effect of Petrochemical Structure on Methane Fermentation Toxicity." Prog. Wat. Tech. 10:545-558.

McCarty, P.L. 1964. "Anaerobic Waste Treatment Fundamentals, Parts I-IV." Pub. Works. 95-197-112.

McCarty, P.L. 1966. "Kinetics of Waste Acclimation in Anaerobic Treatment." Developments in Industrial Microbiology 7, 144. Inst. Biol. Sci., Washington, D.C.

McCarty, P.L., Jeris,. J.S. and Murdoch, W. 1963. "Individual Volatile Acids in Anaerobic Treatment." J. Wat. Poll. Con. Fed. 35:1501-1515.

McCarty, P.L. and McKinney, R.E. 1961. "Salt Toxicity in Anaerobic Digestion." J. Wat. Poll. Con. Fed. 33:399-415.

Miller, T.L. and Wolin, M.J. 1974. "A Serum Bottle Technique for Cultivating Obligate Anaerobes." App. Microbiol. 27:985-987.

Mosey, F.E. 1976. "Assessment of the Maximum Concentration of Heavy Metals in Crude Sewage Which will not Inhibit the Anaerobic Digestion of Sludge." Wat. Poll. Con. 75:10-20.

Neufeld, R.D., Mack, J.D. and Stratkey, J.P. 1980. "Anaerobic Phenol Biokinetics." J. Wat. Poll. Con. Fed. 52:2367-2377.

Neufeld, R.D. and Valiknac, T. 1979. "Inhibition of Phenol Biodegradation by Thiocyanate. J. Wat. Poll. Con. Fed. 51:2283-2290.

Nottingham, P.M. and Hungate, R.E. 1969. "Methanogenic Fermentation of Benzoate." J. Bact. 98:1170-1172.

Owen, W.F., Stuckey, D.C., Healy, J.B., Young, L.Y. and McCarty, P.L. 1979. "Bioassay for Monitoring Biochemical Methane Potential and Anaerobic Toxicity." Wat. Res. 13:485-492.

Parkin, G.F. and Kocher, W.M. 1980a. "Microbial Methane Fermentation Kinetics for Toxicant Exposure." Prepared for the National Science Foundation Water Resources, Urban and Environmental Engineering Program, July 1, 1980, Philadelphia, Pennsylvania.

Parkin, G.F. and Kocher, W.M. 1980b. "Microbial Fermentation Kinetics for Toxicant Exposure." Prepared for the Air Force Office of Scientific Research, September 1, 1980. Philadelphia, Pennsylvania.

Parkin, G.F., Speece, R.E., Yang, J. and Kocher, W.M. 1980. "Response of Methane Fermentation to Industrial Toxicants." Paper presented at the 53rd Annual Water Pollution Control Federation Conference, Sept. 1980, Las Vegas, Nevada.

Parkin, G.F. and Speece, R.E. 1980. "Modeling Toxicity in Methane Fermentation Systems." Paper presented at the ASCE National Conference on Environmental Engineering, July, 1980, New York.

Pearson, F., Shiun-Chung, C. and Gautier, M. 1980. "Toxic Inhibition of Anaerobic Biodegradation." J. Wat. Poll. Con. Fed. 52:472-482.

Regan, T.M. and Peters, M.M. 1970. "Heavy Metals in Digesters: Failure and Cure." J. Wat. Con. Fed. 42:1832-1839.

Rudgal, H.T. 1946. "Effects of Copper-Bearing Wastes on Sludge Digestion." Sew. Works J. 18:1130-1137.

Stuckey, D.C., Owen, W.F., Parkin, G.F. and McCarty, P.L. 1980. "Anaerobic Toxicity Evaluation and Semi-Continuous Assays." J. Wat. Poll. Con. Fed. 52:720-729.

Van Den Berg, L., Lentz, C.P., Athey, R.J. and Rooke, E.A. 1974. "Assessment of Methanogenic Activity in Anaerobic Digestion: Apparatus and Method." Biotech. Bioeng. 16:1459-1469.

Yang, J., Speece, R.E., Parkin, G.F., Gossett, J. and Kocher, W. 1980. "The Response of Methane Fermentation to Cyanide and Chloroform." Prog. Wat. Tech. 12:977-989.

Young, J.C. and McCarty, P.L. 1980. "The Anaerobic Filter for Waste Treatment." Technical Report No. 87, Dept. of Civil Engineering, Stanford University, Palo Alto, California.

CHAPTER 4

MICROBIAL SULFUR RECOVERY-AN EXAMPLE

INTRODUCTION

Problems with sulfur disposal are universal. They are related primarily to energy production and, to a smaller extent, to the mining industry. As more emphasis is placed on the use of hydrocarbons such as oil and coal for energy production, sulfur control and disposal will become a more pressing problem.

One form of sulfur residue is through the generation of $CaSO_4$ sludges, which are a result of the scrubbing of flue gases from coal fired plants. In addition, H_2S waste streams are generated from the desulfurization of coal and oil. This problem is generally handled by using physical-chemical processes (e.g., Claus and Wellman-Lord processes), which convert the H_2S to elemental sulfur. Other, more limited, areas in which high sulfates (acidic wastes) are a problem, include acid mine drainage (coal mining), acid wastes from mining operations, and wastes from ethanol distillaries.

Currently, H_2S waste streams are handled by physical-chemical processes, with the major objective being the recovery of elemental sulfur. Presently, high sulfate wastewaters are neutralized and are either discharged or stored in lagoons.

The biological conversion of high sulfate wastes or hydrogen sulfide waste streams to the more desirable form of elemental sulfur has not been developed beyond the laboratory stage. Studies by Cork (1978), and Cork and Cusanovich (1978, 1979), on the kinetics of sulfate reduction and on the kinetics of hydrogen sulfide conversion to elemental sulfur are, to date, the only published material on the applied bio-process production of elemental sulfur. These studies will be discussed in more detail in a subsequent section of this chapter. In the next section, a brief summary of the microbiology of sulfur transformations will be presented. A more detailed summary has been conducted by Cork (1978).

MICROBIOLOGY OF SULFUR TRANSFORMATIONS

All organisms require sulfur, with the major need in the incorporation of sulfur in proteins. In addition, microorganisms can use sulfur in a similar manner that organisms use the various forms of nitrogen; that is as both an energy source and as an electron acceptor. The various oxidation states and the forms that sulfur exists in the environment are illustrated in Figure 4.1, which represents the sulfur cycle. Of particular interest is the anaerobic portion of the cycle, the reduction of sulfate to hydrogen sulfide, and the oxidation of sulfides to elemental sulfur.

There are a wide variety of bacteria which can utilize sulfate as an electron acceptor in the oxidation of organic carbon. Of particular interest is the genus *Desulfovibrio* which is the group of bacteria most widely mentioned in studies of sulfate reducing bacteria in both natural waters and wastewater. Cork (1978) in his studies on the kinetics of sulfate reduction used *Desulfovibrio desulfuricans*. This particular organism can use a variety of organic compounds including lactate, pyruvate and ethanol. *D. desulfuricans* oxidizes lactate to CO_2 and acetate while reducing sulfate to hydrogen sulfide. This process is expressed as follows:

$$\underset{\text{(lactate)}}{2CH_3CHOCOO^-} + H_2SO_4 \rightarrow \underset{\text{(acetate)}}{2CH_3COO^-} + 2HCO_3^- + H_2S \quad (4.1)$$

However, in studies conducted by Middleton and Lawrence (1977) in which a mixed culture of sulfate reducing bacteria was used, the substrate was acetate. Furthermore, complete oxidation of acetate to CO_2 was achieved in contrast to the partial oxidation of lactate to CO_2 by a pure culture of *D. desulfuricans*. The implications of these differences in terms of treatment processes will be pointed out in a later section. In both studies the sulfate reducing bacteria were found to be relatively

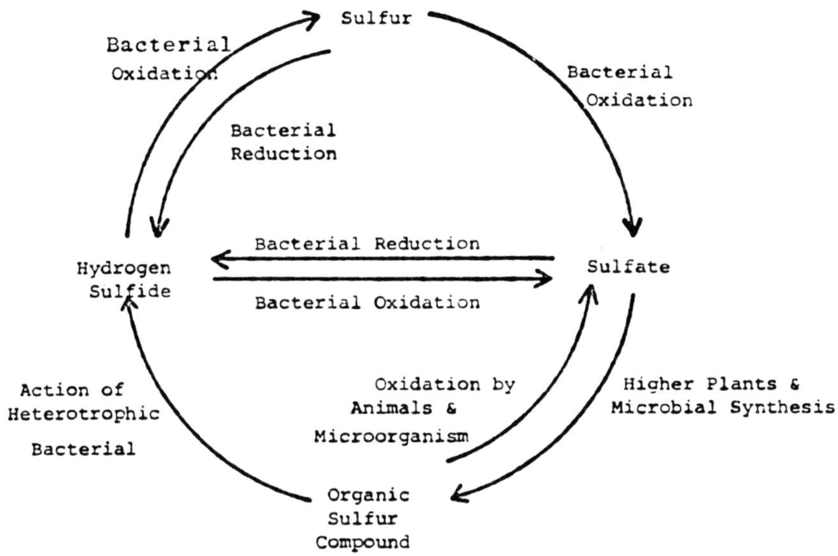

Figure 4.1 The Sulfur Cycle (Cork, 1978).

fast growing. At 30°C the D. desulfuricans species was found to have a maximum specific growth rate of 1.0 day^{-1} [Cork, 1978], while the mixed culture had an estimated maximum specific growth rate of 0.55 day^{-1} [Middleton and Lawrence, 1977], or roughly 50 percent of the pure culture. In addition, the cell yield for the mixed culture was found to be low, at 0.065 mg/mg acetate. Interestingly, the endogenous coefficient was calculated to be 0.0 day^{-1}.

In the biological conversion of sulfide to elemental sulfur, the key microorganism is an anaerobic sulfur bacterium which converts H_2S to S^o, photosynthetically. Hydrogen sulfide acts in a similar manner as water does in plant photosynthesis, in acting as a hydrogen donor for the subsequent fixation of inorganic carbon. Surprisingly, the reaction occurs in sediments of shallow lakes, with the action spectrum of the reaction center chlorophyll being 870 nm, which is out of the visible light region and closer to the infrared region [Cork, 1978]. The overall equation linking bacterial photosynthesis to sulfide oxidation is as follows:

$$2H_2S + CO_2 \xrightarrow{\text{Light}} \{CH_2O\} + S^o + H_2O \qquad (4.2)$$

In addition, studies by Sirevag and Ormerod (1977) have shown that significant amounts of the carbon synthesized can be stored in the cell, as polyglucose. Furthermore, in the absence of light (i.e., in a dark reaction), the polyglucose was utilized and three fatty acids were excreted, with 80 percent being acetate. The capability of these organisms to both oxidize hydrogen sulfide to elemental sulfur and subsequently fix carbon (polyglucose) which can be degraded in a dark reaction to primarily acetate, leads to several interesting sulfur recovery process possibilities. These proposed processes will be described in more detail in a later section.

The specific growth rates of Chlorobium thiosulfatophilum were found to be relatively high. At 30°C, Cork (1978) and Cork and Cusanovich (1979) found that the maximum specific growth rate of C. thiosulfatophilum was 1.0 day^{-1}, which is comparable to heterotrophic organisms. Thus, it does not appear that the specific growth rate of the bacteria would be an important factor, as is the case of some _Methanobacteria_ in mixed culture which have considerably lower growth rates [Middleton and Lawrence, 1977].

INDUSTRIAL APPLICATIONS

Presently, the use of bio-processes for sulfur production or recovery have been limited to laboratory pure and mixed culture studies. As mentioned previously, Cork (1978) and Cork and Cusanovich (1978, 1979) have proposed a gas purged mutualism which utilizes _Desulfovibrio desulfuricans_ and _Chlorobium thiosulfatophilum_ in the treatment of a high sulfate metallurgical effluent. Figure 4.2 illustrates the basics of the process, i.e., the sulfate reduction in the first reactor followed by hydrogen sulfide oxidation in the second reactor. The hydrogen sulfide is purged from the first and bubbled through the second, using an inert carrier gas. They felt that one of the limitations of the process was the (lactic acid) organic carbon required to drive the sulfate reduction. In addition, their work is the only reported work in which _Chlorobium_ was studied with the specific objective of elemental sulfur recovery.

Middleton and Lawrence (1977) studied the kinetics of sulfate reduction, using a mixed culture of sulfate reducers, with acetic acid as substrate. They proposed a scheme (see Figure 4.3) in which wastes from SO_2 scrubbers and/or sulfuric acid wastes could be treated. The high sulfates from both sources could be removed by sulfate reduction to hydrogen sulfide. In addition, the sulfide and alkalinity produced could be used for precipitation and neutralization of an acid waste. Again, the limitation of the process would be a requirement for organic carbon substrate. The sulfur recovery system which they proposed was a chemical-physical one. In addition, optimization of the sulfate reduction step would be required and could be done in a similar manner in which nitrate removal is optimized, i.e., by using a fluidized bed reactor or a sludge blanket reactor. In a similar manner Pugsley, _et al_. (1971) proposed the use of a sulfate reduction process in the treatment of mine drainage in Colorado. Again, the limitation appeared to be in the organic carbon requirements.

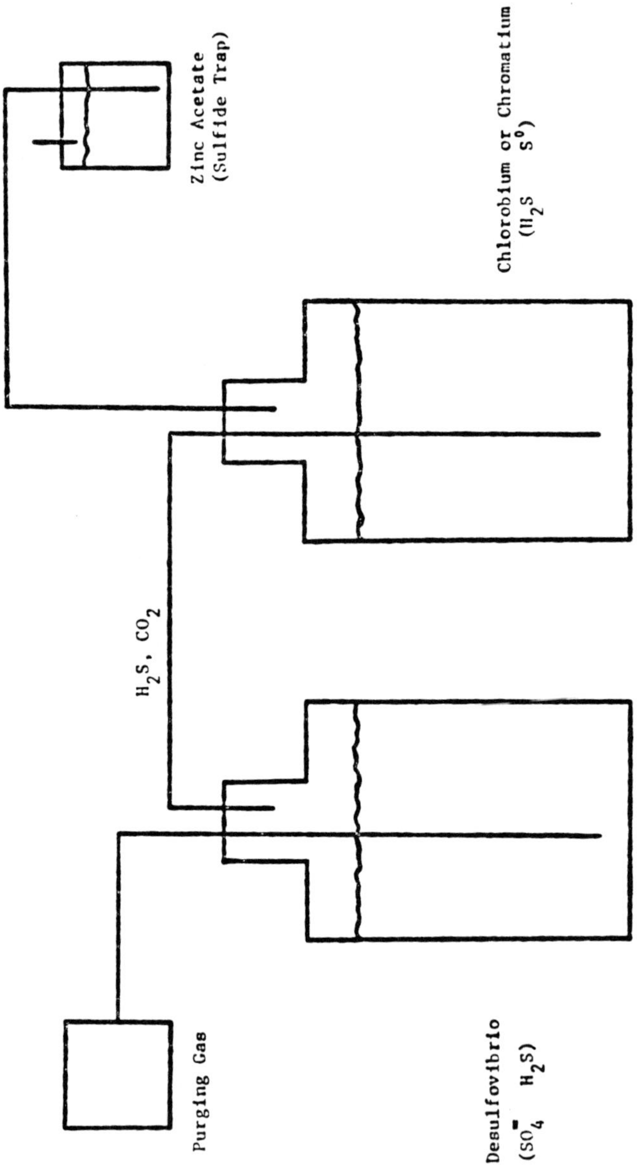

Figure 4.2 The Gas Purged Mutualistic System Utilizing Desulfovibrio and Either Chlorobium or Chromatium (Cork and Cusanovich, 1979)

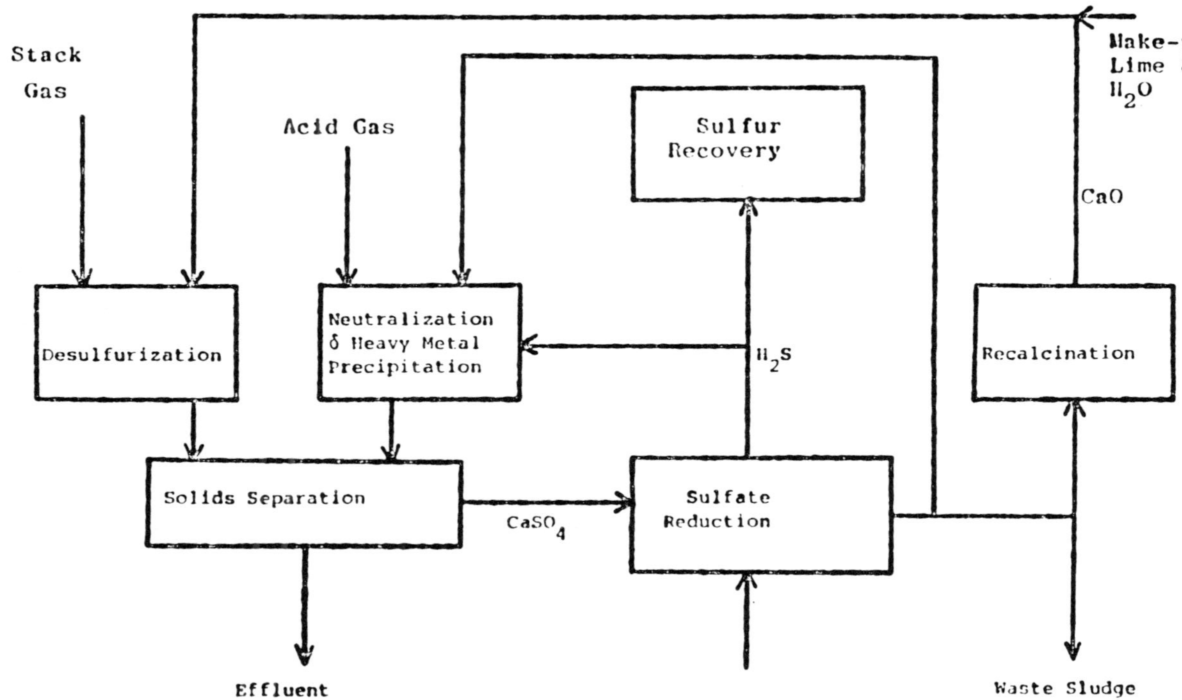

Figure 4.3 Possible Process Schemes Using Microbial Sulfate Reduction for Treatment of Sulfuric Acid Wastes or Reclamation of Calcium Sulfate Sludges (Middleton and Lawrence, 1977).

EXAMPLE PROCESS TREATMENT SCHEMES

The following two process schemes were developed in collaboration with Dr. Douglas J. Cork, Department of Biology, Illinois Institute of Technology. As shown in Figure 4.4, the first schematic depicts the proposed process scheme for treatment of a high sulfate wastewater. The two basic unit processes are the sulfate reduction reactor and a photolithotrophic fermentor which is similar to the process used by Cork (1978). One major difference is the use of the photolithotrophic fermentor to provide the organic material required to drive the sulfate reduction step. The essential feature of the process is the recirculation of enough organic carbon to reduce the sulfate in the wastewater. Another major difference is the use of dark reactors in order to induce the conversion of synthesized and stored polyglucose to acetic acid. Physical processes would be required for the separation of the sulfur and acetate from the bacteria, and subsequent recycle of biomass back to the fermentor.

The second proposed system involves the biological (see Figure 4.5) treatment of H_2S waste gas streams which are normally found in the desulfurization of hydrocarbons and coal gasification [Cork, 1982]. In this case, the sulfate reduction step would not be required and the organic carbon (acetate) from the dark reactor could then be channeled into an anaerobic reactor for the recovery of methane, a potential energy source. Thus, light energy could be directly converted into a readily usable form of energy (methane), while oxidizing H_2S into a elemental sulfur. This would have an advantage over the Claus process, in which methane is not recovered.

The process schemes cover the majority of pollution problems associated with sulfur, i.e., high sulfates and H_2S, and can potentially convert them to elemental sulfur, the most commercially desired form of sulfur. In addition, energy recovery is possible. Both processes use anaerobic microorganisms and thus have the advantages of no aeration equipment required, low sludge production, and the processes appear not to be energy intensive.

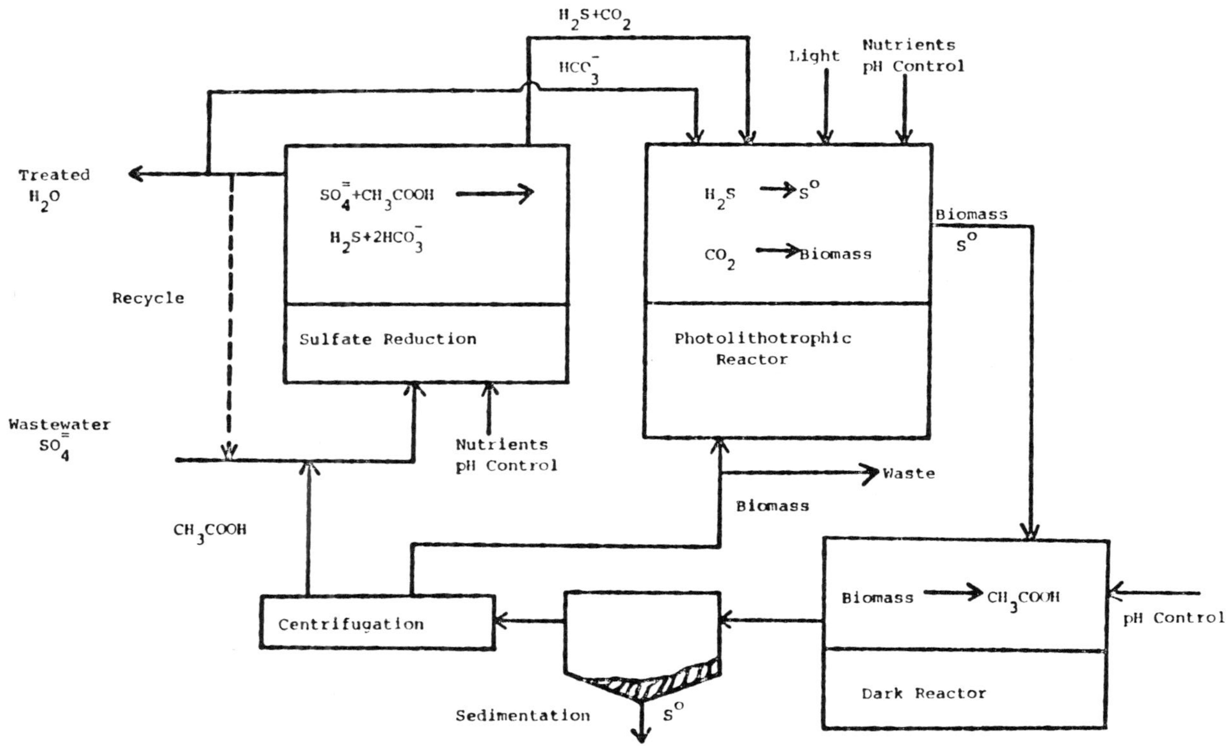

Figure 4.4 Integration of Anaerobic Processes for Sulfur Recovery From a High Sulfate Wastewater

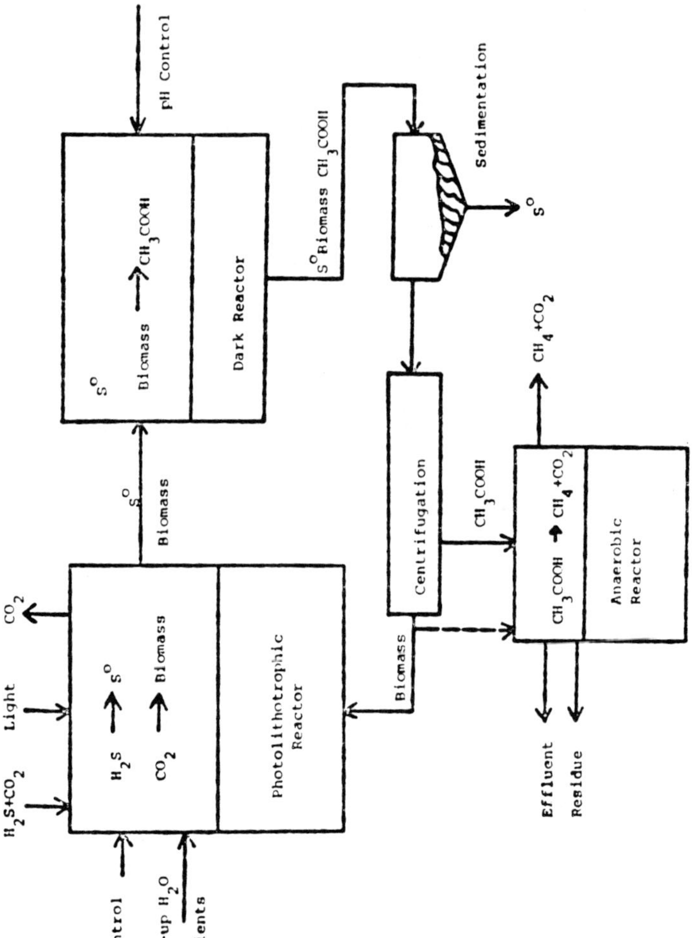

Figure 4.5 Integration of Anaerobic Processes for Sulfur and Energy Recovery From an Acid Gas Stream

Because of the impact both processes would have on controlling sulfur in the environment, there appears to be justification for fundamental research in two major areas. In the first area, studies should be conducted on the Chlorobium species to determine such things as:

1. Maximum Overall Substrate Utilization Rate (Optimum organism concentration)

2. Minimum Light Requirements

3. Kinetics of Acetate Conversion in the Dark Reactor

4. Factors Affecting the Storage of Polyglucose

A second area of research would involve the sulfate reducing step and would include such things as:

1. Optimization of the Sulfate Reduction Process

2. Toxicity Problems and Control Strategies

SUMMARY

The control of sulfur pollution is a widespread problem and mainly associated with hydrocarbon processing and power production. For example, the mining of coal results in acid mine drainage, the burning of coal results in $CaSO_4$ sludges (SO_2 scrubbing), and coal and oil desulfurization results in H_2S waste streams. Of the two forms of sulfur pollution SO_4^{-2} is a major problem with no reliable method of control.

A review of the literature indicates that microbial sulfur control is only in the formative stages. There is also an indication that potential exists for the use of both the sulfate reducing bacteria and a photolithotrophic organism Chlorobium which oxidizes H_2S to S^o using light. Two process schemes were proposed for both types of sulfur problems, i.e., high sulfate wastewaters and H_2S gas streams. The process scheme for the treatment of high sulfate waters is particularly promising since the normal requirement for an outside source of organic carbon for sulfate reduction would not be required. However, before process development can begin, work needs to be accomplished on the photolithotrophic bacteria, *Chlorobium* and the sulfate reducing bacteria.

REFERENCES

Cork, D.J. 1978. *Kinetics of Sulfate Conversion to Elemental Sulfur by a Bacterial Mutualism: A Hydrometallurgical Application*. Ph.D. Thesis, University of Arizona.

Cork, J.D. 1981. "Design and Utilization of a Bench Scale Gaslift Fermentor." *Dev. Ind. Microbiol*. 22:733-738.

Cork, J.D. 1982. "Acid Waste Gas Bioconversion - An Alternative to the Claus Process." *Dev. Ind. Microbiol*. 23:In Press.

Cork, D.J. and Cusanovich, M.A. 1978. "Sulfate Decomposition: A Microbiological Process." In *Applications of Bacterial Leaching and Related Microbiological Phenomena*, Ed. L.E. Murr, pp. 207-222. Academic Press, Inc., New York.

Cork, D.J. and Cusanovich, M.A. 1979. "Continuous Disposal of Sulfate by a Bacterial Mutualism." *Dev. Ind. Microbiol*. 20:591-602.

Middleton, A.C. and Lawrence, A.W. 1977. "Kinetics of Microbial Sulfate Reduction." *J. Wat. Poll. Con. Fed*. 49:1659-1670.

Pugsley, E.B., Cheng, C.Y., Updegraff, D.M. and Ross, L.W. 1971. "Removal of Heavy Metals From Mine Drainage in Colorado by Precipitation." *Chem. Eng. Ser*. 67:75-89.

Sirevag, R. and Ormerod, J.G. 1977. "Synthesis, Storage and Degradation of Polyglucose in Chlorobium Thiosulfatophilum." *Arch. Microbiol*. 111:239-244.

INDEX

Acetic acid, 23, 126, 157
Activated sludge test, 53
  advantages, 54
  disadvantages, 52
Ammonium, 28
Anaerobic contact processes, 9
Anaerobic degradation, 1, 170
Anaerobic digesters, 116, 123, 136
Anaerobic digestion, 26
Anaerobic filter process, 11, 12
Anaerobic rotating biological contactor (ANRBC), 21
Anaerobic systems, 86
Anaerobic upflow filter process, 138, 145, 174
Anaerobic waste treatment, 3, 5, 9, 11, 109
Aromatic compounds, 57,
  anaerobic photometabolism, 57
  methanogenic fermentation, 58
  biochemistry, 63

Batch toxicity tests, 124, 125, 131, 132, 143
  evaluation of, 140
Biodegradation, 47
  primary, 48
  ultimate, 48
  acceptable, 48

Calcium, 28
Catechol, 66, 67
Celanese Company, 11
Complex formation, 116
Compounds,
  metallic and organic, 171
Continuous assay, 138
Continuous toxicity tests, 145, 146
Copper, 116, 174
Cyanuric acid, 75

Digestion process, 36

Expanded bed process, 12, 14

Flask test method,
  advantages, 53
  disadvantages, 53
Formaldehyde, 115, 158, 175

Gas production, 126, 128
Gas production rate, 18
General Electric, 87
Genetic engineering, 86, 88

Heavy metals, 116, 118
High sulfate wastes, 179

Hydrocarbons, 84
Hydrogen sulfide
182
Hydrogen sulfide
wastestreams, 179

Inhibition, toxicity,
antagonism, synergism,
110, 115

Magnesium, 28
Maize-starch waste, 21
Methanogenic bacteria, 7,
23
Microbial growth, 118
Modified clarigester, 19

Nickel, 120, 121, 126,
128, 140, 174
Nitrogen, 28, 29

o-cresol reactor, 67
Organic compounds,
removal of, 122

Paraffin oil, 84
Paraffin wax, 84
Pesticides, 77
lindane, 78 80
DDT, 79
aldrin, 80
endrin, 80
heptachlor, 80
Petrochemical industry,
115
pH and alkalinity, 23, 26,
Phenol, 66
Phosphorus, 29
Polyglucose, 182
Process treatment schemes,
186

Reversibility toxicity
test, 146
River die away test, 50
advantages, 50
disadvantages, 51

Semi-continuous
control method,
134
Semi-continuous
serum bottle
assay, 133
Semi-continuous
toxicity test,
144, 145
Sludge bed zone, 15,
16, 18
Sludge blanket
section, 16, 18
Soaps and
detergents, 73
Sodium acetate, 126
Sulfur disposal, 179
Sulfur pollution,
189
Sulfur recovery, 184
Sulfur
transformations,
180

Temperature effects,
39
Toxicity assays,
124, 130
Toxicity effects,
107, 109, 112,
136
Toxicity models
empirically
based, 147, 169
kinetically
based, 153, 169
inhibition
coefficient, 163
Trisodium carboxy-
methloxysuccinl,
ate, 75

Upflow anaerobic
sludge blanket
process, 15, 19,
37

Warburg
Respirometer, 51, 129
advantages, 52
disadvantages, 52
Wastes
sugar processing, 68
food processing and agricultural, 68
brewery, 69
animal, 69
pharmaceutical, 76
sugar beet, 18, 19

Zinc, 116, 174